D0890585

. . . . a movement of one . . . .

The future depends on what we do in the present

. . . . Mahatma Gandhi

written & illustrated
by Dana Simson

cover illustration . love is symmetrical

dedicated to
Jackie Fritch
celebrating her life-long work
for civil rights & peace

Printed in the United States

The paper used in this publication is produced by mills committed to responsible and sustainable forestry practices.

1 2 3 4 5 6 7 8 9 01

Green Writers Press is a Vermont-based publisher whose mission is to spread a message of hope and renewal through the words and images we publish. We will adhere to our commitment to preserving and protecting the natural resources of the earth. To that end, a percentage of our proceeds will be donated to social-justice and environmental activist groups. Green Writers Press gratefully acknowledges the generosity of individual donors, friends, and readers to help support the environment and our publishing initiative. For information about funding or getting involved in our publishing program, contact Green Writers Press.

Giving Voice to Writers & Artists Who Will Make the World a Better Place

Green Writers Press | Brattleboro, Vermont
greenwriterspress.com

ISBN: 978-1-9505840-5-5

. . . y o u   a r e   t h e   c h a n g e . . . . .

a handbook to help you

rethink to reduce your use of harmful products & practices

retool your home & lifestyle

reduce . reuse . recycle .

restore your health

refill with joy and optimism

return to a simpler, stress-free self

reduce clutter & distraction

save money

save the planet

i n d i v i d u a l   a c t i o n   o n   c l i m a t e   c h a n g e

. . . . a movement of one . . .

t a b l e o f c o n t e n t s

how to use this book

Feel free to read it through or flip it open randomly,
get ideas and build on them,
or use it as a reference.
Start with what you can
easily change.
All actions taken to
retool your lifestyle
get a gold star
for effort.

Discover simple ways for individuals to stop contributing to climate change today, and start bottom-up positive change for a healthier planet for all living things.

what we did

There is a common goal that can unite all people on earth, beyond differences in politics, race, religion, culture, or disbelief:

to address climate change
and human impact
on earth's ecosystems.

Who am I to leap up on a soapbox beating a drum? Just another one of the 7.6 billion people over-using our planet's finite resources with little regard to future generations. As someone with a deep love for nature and all of Earth's creatures, in the absence of a spokes-animal, I have also volunteered to stand up for them.

Historically, humans have been inclined to be self-focused on having more, and better. We have invented, built and amassed amazing things during our short upward ride.

Today's new technologies, conveniences and creature comforts are at an all-time high, but an uneasiness persists that we may be standing on some very thin ice.

The cost of centuries with little regard for earth's basic ecosystems has resulted in placing animal species at risk, contaminating our air and water, heating up the atmosphere and seas, crossing a tipping point beyond which without immediate action we will not survive.

what we do now

. one .

We keep waiting for our leaders to fix things. Sorry. They won't. They are in office for a short time so focus on short-term issues to stay in their position of power.

Maybe corporations will care enough to self regulate and invent earth-friendly products. Some do. Most won't. Capitalism has slid from pride in product to a me-first attitude that slyly justifies dirty practices to make a buck.

Ah, but that is where you, dear reader, discover you are far stronger than your shoe size implies. When one of us retools his or her thoughts to live mindfully for the future, those decisions may hit a switch in the small circle of people that surround you. They will see you at the restaurant, when offered a polystyrene box for your extra chicken wings, ask pleasantly instead for several sheets of wax paper. Their eyebrows may lift slightly to see you easily wrap up those lunch snacks and leave in a small green victory. Small earth friendly actions paired with a new mindfulness will raise important questions about the logic of single-use plastics, polar level AC in grocery stores or our continued addiction to fossil fuels. You, gentle warrior, are the decider of your next green action.

There is no appeal for money, votes, emails, or lengthy meetings here. This is a handbook for anyone to simply rethink their daily actions for a better tomorrow. A future that will be determined by how each of us acts today.

. another .

ideas & action

Instead of waiting for top down action, it is possible that one by one, through this grassroots movement, we can become empowerd to address climate change on an individual level.

one may become many

When you need a little help, you get by with the help of your friends. For added insight in the writing of this book I enlisted a group of thinkers, old and new friends, to prod me on with their questions and observations. I hope you will have some great thoughts too. Ideas represent our ace in the hole, along with creativity and invention, basically tapping the same smarts that got us into this hot mess.

we are one

Global warming is a deeply complex issue, already affecting many aspects of our planet. This slim book will just skim the surface. I intend it as a call to arms, something to refer to on your journey forward in green awareness and action.

The title suggests an action:

. Come Together .

If we are to succeed in finding a solution to climate change, we must engage one another. The divisive and unsettling time we find ourselves in requires a section exploring appeasement and respectful communication along with individual action.

Let's sit down
without animosity to find a way to link arms.
The prize is great:
restoring health to life on earth for future generations.

There are and will be further migrations of people due to climate change, war and economic strife. We need to learn how to accommodate one another to live together peacefully.

so let's talk

ideas & action

We have lived our lives by the assumption that
what was good for us would be good for the world.
We have been wrong.
We must change our lives
so that it will be possible to live by
the contrary assumption,
that what is good for the world will be good for us.
And that requires that we make the effort
to know the world and learn what is good for it.

.... Wendell Berry

The Thinkers Gang

John Orth

Vickere Murphy

Heidi Thompson

Joey Simson

Pam South

Beverly Bevan

Mary Kate Bland

Chere Petty

Pat Russel

Lisa Jo Frech

April Chapman

Bobbie & Randy Stadler

much appreciated

crossing the divide

..........connecting across division & mistrust

tools .

constructive listening

empathetic respect

the art of conversation

whole earth dialogue

The divisive tribalism that has arisen like putrid smoke to cloud our senses and fog logic is contagious. Negative energy can be a match firing up mistrust and righteous indignation. This primal uproar may be based deep within each of us from back when we guarded our scrubby caves from intruders. Once someone stands their ground, any attempts to question or disarm can feel like an assault.

To move ahead into civility requires simple tools. Remaining calm works better to smother a squabble than picking up a stick.

Offering respect and listening without judgement may help to set another who is on edge, at ease.

Many of the tools I offer here are already in each of our toolboxes. Like any tool that is increasingly employed, more use will result in making it a natural go-to. Cultural swings in social behavior set new norms for interactions between people.

Right now, we could be nicer.

**I'm a very strong believer
in listening and
learning from others.**

. . . . Ruth Bader Ginsburg

Listen up.
Dictating a command,
talking past or over others,
stating your position as fact,
attempting to win an exchange . . . .
Puts others in a defensive position.

Multi-task listening, self-absorbed listening,
or a focus on gathering your response. . . .
Means you are not listening.

A difference of opinion
does not necessarily mean
someone is wrong.

. explore .

Why do they have this notion?

Can you learn something new or
better understand an alternative view?

Is it possible to engage this person
in a positive way and agree to disagree . . . .
and move ahead together?

10 ways to improve listening skills

1 . Open your heart

Existentially each of us is bumbling through our lives without a map, so why not cut yourself and others some slack.

2 . Fully engage one another

Face each other. take time to make eye contact and truly listen.

3 . Be present & relaxed

Consider what is being said. Quiet any distractions or your own thoughts, feelings or biases. Unplug ego.

4 . Keep an open mind

Try not to judge even if the speaker says things that alarm or annoy you. Keep in mind that words may reflect someone's life experience, thoughts or feelings. In listening, you may gain a more complete idea of why they think like they do.

5 . Imagine what is being said

Put yourself in the shoes of the person you are listening to. Feel the words from their perspective. Listen with empathy.

6 . Don't interrupt

People think and speak at different rates. If you are good at expressing yourself, try not to jump in if someone is slower at communicating. Interrupting signals that you don't really care what the other person is saying.

7 . Ask questions to clarify

Wait for a pause in the conversation, if you wish to gain a better understanding of something, then refer back to the statement in question.

8 . Give positive feedback

While you don't have to agree, you can draw people out by showing you are interested in the speaker's point of view.

9 . Note what isn't said

You can learn a lot by noting what is left out of a conversation. People's faces sometimes convey far more than what they actually say.

10 . Briefly summarize your understanding at the end of a conversation

Make sure you both are on the same page. Frequently the same sentence can be heard differently by two people. Summarizing also verifies you cared to listen.

Follow the three R's
Respect for self
Respect for others
Responsibility for
all of your actions.

... Dalai Lama

Respect is underrated as a tool in good communication. In the past it formed the very basis for how one would speak to another. Currently on TV, in public or even with friends or family, people insult or interrupt, loudly talking past each other without really listening.

Respect allows for polite interaction by all parties.

Sometimes we have to deal with pretty sketchy people at our gallery. Anyone can walk in the front door: drunks, vagrants, crooks or angry mentally unstable people. It's important to greet them as you would any customer and shower them with attention to keep them on an even keel until they leave. A positive approach does a good job of keeping bad behavior at bay.

Empathetic respect is feeling where the other person is coming from.

Reserve judgment

A bad first impression may lead to a disrespectful exchange that is unwarranted. One person may assume that they are more knowledgeable than another, speaking down to them. This behavior is very off-putting. I have witnessed this degrading posture from doctor to patient, men to women, across races, ethnicities, age, political stances, economic standing and stature.

Why bring someone down?

A disrespectful person loses respect from others immediately.

If your goal is to convince someone of something or have a productive dialogue, work to keep the playing field level and welcoming.

There is a difference between an argument and a discussion, that is, whether you feel angry or enriched afterward.

We need to make sure we're all working together to change mindsets, to change attitudes, and to fight against the bad habits that we have as a society.

. . . . Justin Trudeau

Ideal conversation must be an exchange of thought, and not, as many of those who worry most about their shortcomings believe, an eloquent exhibition of wit or oratory.

.... Emily Post

Social media, email, and even phone calls have undermined what has been called the art of conversation. We converse in short sound bites, rarely taking time to truly connect.

con·ver·sa·tion . noun
the informal exchange of ideas by spoken words.
"the two men were deep in conversation"
Synonyms: discussion, talk, chat, gossip, tête-à-tête, heart-to-heart, exchange, dialogue

Ideas and shared consensus are born through conversation. Difficult issues like global warming must be discussed to find useful solutions. Talking together about our concerns, exploring positive action or taking steps toward collective goals must start with an engaged conversation that unites us.

As each one of us chooses an environmentally responsible path we can respectfully welcome others along on the journey.

A great conversation
is like playing a fun ping-pong game
without keeping score.

1. Reach out
Start a conversation by finding shared experiences. It may start with a compliment, observation, or comment on something as simple as the weather.

2. Take the conversation deeper by focusing on a common interest or topic you're both interested in.

3. Share the floor, equal contributions from both parties allow each speaker to feel listened to and engaged.

4. Ask open-ended questions that encourage a deeper understanding.

5. Be polite. Don't impose your view or argue with the other person.

6. Be mindful of the other person's time. A good conversation is a gift of time and attention.

**Some may still deny
the overwhelming judgment of science,
but none can avoid the
devastating impact of raging fires
and crippling drought and
more powerful storms.**

.... Barack Obama

Even though we find ourselves undermined by climate deniers in places of power determined to roll back important environmental advances, local and state governments, as well as, much of the global community are taking measures to stem climate change.

There are thousands of organizations big and small, worldwide, working on various aspects of this important issue. Their focus ranges from global warming's effects on animals, cities, agribusiness, political/social implications, sea level rise & acidity, to extreme weather and increasing temperatures globally. Scientists and inventors are developing exciting solutions (visit Drawdown.org). Sustainable practices used for centuries are being revisited, such as rainwater collection systems, no till farming and a trend back to organics.

The issue of global warming offers us an important opportunity to rethink and retool the way we have been doing things. Instead of paralysis or despair, this can be a creative time of working together, signaling a return to living responsibly for long-term gain for everyone.

Don't Drink the Kool-Aid

The toxic din of real or fake news, hearsay and chatter can be quieted. It is easy to feel overwhelmed by conditions we may face as the earth warms. This stress may actually contribute to deniers, general ignorance or the problem of motivating action to make the changes we need to, both personally and globally.

Small steps taken as an individual are helpful in securing some control over our planet's future.

a movement of one

Imagine if the small actions we discuss in the next chapters become a mindful lifestyle adopted by you, then your friends, your family, next, their circles of contacts, and building outward. Happily these simple changes will also save you money, lessen daily stress, reduce clutter and fill you with an empowered joy of doing something proactive.

We live in a world of excess. It barely scratches the surface of our comfort zone to knock things back a few notches. So, let's get started...

living for the future

.........bottom up change to slow global warming

tools .

finding a new comfort zone

proactive convenience

mining the past

enough already

When you start your day tomorrow, focus on noting areas of excess. Grab a pencil and pad to list things you can change right now in your home, at work, and throughout your immediate environment. Lists are great because as you address each thing you get to tick it off, which is a wonderfully fulfilling act.

Channel your dad in the 1970s
Or if your dad didn't do this,
I will let you channel my dad; Bruce.

There was a big recession back then with gas lines and lots of penny pinching. We hand-washed & air-dried the dishes. When you left a room the lights had to be turned off, pronto! In winter, window shades went up to catch heat during the day and down to retain warmth inside at night. Showers were short. Water was conserved. Clothing was handed down. People carpooled. We ate little meat. Soda cans and bottles were always picked up because you got a nickel back for each one you returned. The house heat was kept low and we piled on sweaters. In summer, instead of AC, we just opened the windows, ate Popsicles, and ran through the sprinkler when it was hot. Clothing dried on the line. The washer was run only when it was full. Paper bags and refillable glass milk bottles were the norm. Straws and most containers were made of paper.

a new comfort zone

Everyone used far less power. People read or played board games instead of watching hours of TV or hanging on their computers. As teenagers, a bike was our primary mode of transportation and most families got by with one car. We drank water from the faucet. I could go on, but you get the picture.

Our current comfort zone has risen to wasteful levels.
NOTE: a single leaf blower in continuous use for an hour produces the same amount of pollution as seventeen cars.

Today we indulge in big cars, parades of energy sucking appliances, ridiculous trends like mail-order meals (with each ingredient packed in individual plastic bags), night assaulting residential yard lighting, heated swimming pools and garages, bottled water, mountains of throwaway single-use containers and piles of cheap products made thousands of miles away. Electronics must be replaced frequently with the latest model. There is widespread use of pesticides that clear weeds but also contaminate aquifers and kill honeybees. People use deafening leaf blowers to blast leaves into the street or a neighbor's yard instead of just raking them up.

Comparatively our current comfort level borders on ridiculous wretched excess or at least collective absurdity.
So enjoy making that list.

Where there is too much something is missing.
.... Leo Rosten

a new comfort zone
Dialing back what we use
is a mindset to use or take less.
You may find yourself far more comfortable
with more money in your pocket.

Global warming is the result of many ills.

In changing our habits to lessen the footprint we leave behind, living in an environmentally responsible manner is the first step. We have trusted the products, cultural trends, and decisions marketed to us politically and economically. I admit to rarely reading labels—ok never reading labels in the past. I also click agree, consistently ignoring paragraphs of tiny type of what I agree to on websites. It sure would be great if we could trust those getting our dollar or vote to treat us honestly, but unfortunately that asumption has not served us well. Profits have replaced integrity in manufacturing resulting in a multitude of health issues for living things and our planet.

It's time for each of us to take the wheel on our collective destiny.

The good news is,
we don't have to buy what they are selling.
The pocketbook is power.
Consider what you bring into your home, body and environment.

Let's start where we live.
This is a place where we have total control
and so much we can easily improve.

a new comfort zone

Bathroom

small room big mess

Open the door.
Find an oasis of easy improvements.
From the toxic PVC's lurking in your shower curtain to tossing toiletries, plastic dental, health or beauty products, poor disposal of old vitamins and pharmaceuticals, to using miles of TP and wasting endless gallons of water, small changes can result in big improvement.

Water usage requires lots of energy to process, pump in and then properly dispose of the used gray water and sewage. Using less will save money and reduce environmental impact through mindful conservation.

Turn off the water

while you brush your teeth, shave or soap up.
Turn water back on to rinse.
Cut down on the length of showers
or the amount of water in your bath.
A night toilet means flush once in the morning.
Flush less with #1, kind of yucky, but saves lots of water.
22 gallons of water are flushed down the toilet daily
in an average U.S. household.

Low-flow toilets and faucets cut down on the amount of water used. A quick low-tech solution for older facilities is to put a brick in the back of your toilet tank so it takes less water to fill it again after each flush.

A typical shower head will use 5 to 8 gallons of water per minute; installing a low-flow head, will bring that down to 2.5 gallons with equal water pressure. Take a shower every other day. You can sponge bath with a washcloth in between.

New technologies are being developed to reuse gray water, especially in drought prone areas. Be aware of how much is going down your drain.

the issue with tissue

Paper is undervalued, overused and thrown away.
While CO_2 absorbing trees are a renewable resource, they take years to grow and represent an important element in combating climate change. We need more trees instead of wiping them out.

Avoid using thick *quilted toliet paper; instead purchase thin TP made with 100% post-consumer recycled paper, processed without harmful chlorine and free of BPAs. *80% of TP sold in the U.S. is made by clear-cutting the "Canadian Amazon" boreal forest. There is no recycled content in Procter & Gamble (Charmin), Kimberly-Clark (Scott) or Georgia-Pacific, owned by the infamous Koch Brothers (Angel Soft, Quilted Northern).

buy . better

Scott tube free: no cardboard roll, but packaged in plastic. ($.58 per roll)
Who Gives a Crap: recycled, $1 per roll, free shipping for 48 roll pack. No BPAs, plastic-free, 100 percent recycled, but bleached.
Seventh Generation Unbleached $0.83 per roll (a rougher feel) but is unbleached and 100 percent recycled from a high integrity company.
Renewable bamboo TP is best earth-friendly option, but more pricey at $1.70 a roll.

TP wipes out acres of forest!

27,000 trees a day are flushed away daily worldwide.

quick FIX:

Insert the roll to roll over rather than roll under to use less each time.

The paper industry looks to tree plantations to create an ongoing supply of virgin pulp and fiber, but these monocultures displace indigenous plant and animal life, while requiring tremendous amounts of water, chemical pesticides and fertilizers. There are new tissues made of more renewable materials than wood, some better for septic systems.

In India, researchers have actually harvested TP from sludge, drying it out and using it to make electricity. The energy generated equals the efficiency of a natural gas plant at the same cost as in-home solar panels. Not a job I would want, but very enterprising.

nobody nose

There are "green" facial tissues out there, but they still come in a cardboard box that empties alarmingly quickly. Use your "green" TP instead; it is cheaper & unpackaged or better still, employ an old fashioned washable hanky!

even worse, period

Old school sanitary napkins were mainly made with paper but the newest generation contains absorbing petro-polymers, plastic and bleached wood product. British environmental activist Ella Daish laments, "We use a sanitary pad for four to eight hours, but they take 500 years to disintegrate." In the U.K. 3.4 million tampons and 2.4 million pads get flushed away every day. Some tampons even have plastic applicators that have washed up on beaches as far away as the Maldives.

There are safer, healthier, environmentally savvy alternatives available. Moon cups are affordable, they collect the flow, are reusable and last up to 10 years. Period pants, period-proof underwear, can be substituted for regular underwear during menstruation. Both of these options are far less costly to the pocketbook and earth than the status quo.

what to doodoo about Disposable Diapers

Americans throw away 49 million diapers per day, adding up to 2% of the garbage in landfills. Many disposable diaper brands contain chlorine, leading to skin irritations or a rash, as well as latex, perfumes, or dyes that may trigger allergic reactions. The reduced breathability can create a raised temperature that is unhealthy for male babies. Disposable diapers contain toxins harmful to animals, humans, and the environment (polyethylene, petroleum, wood pulp, gelling material, perfume, polypropylene, and non-renewable, petroleum-based ingredients).
The production of a disposable diaper releases the carcinogen dioxin, also found to pose a threat to developing fetuses. Even the "pluses" of disposable diapers present a problem. The wicking dry component has proven to make toliet training more of a challenge to toddlers that are content to stay in dirty diapers.(Butt) happily at least new biodegradable diaper options are leak proof and chemical free.

Go this route please.

seriously deep doodoo

As our population ages, adult diapers contribute 7% to landfills, where they do not biodegrade. Currently there are no green alternatives other than booster pads or a "diaper doubler" that goes inside a reuseable diaper exterior. In Europe "long-lasting" 8 hour diapers are available.

These are some pretty wild fixes, but ideas are what we have.

The brand EcoBaby has suggested vermicomposting diapers (to basically bury and compost diapers in your backyard). You would have a bunch of super happy worms. A variety of experimental diaper recycling programs are being explored worldwide. Used diapers are collected in special bags, then they are sanitized and disassembled to be put into new items. The exterior diaper material is recycled into a variety of construction materials while the inner pulp material is used for shoes, oil filters and wallpaper. I may look at wallpaper in a whole new way now.

Ok enough potty talk.
I'm going to clean things up now.

greener bath linens

Bamboo has antibacterial qualities when spun into linens. Towels made from 100% bamboo or rayon and bamboo blends are far softer, dry faster, are more absorbant and durable than traditional cotton towels. There are many environmental benefits of fabrics derived from bamboo: its natural abundance worldwide requires no pesticides and little water. Commercial cotton depletes the soil, requiring heavy pesticide and high water usage. If you must use cotton, at least buy organic. Bamboo towels dry faster too, There are great hemp products coming on the market now also, so try to choose earthwise.

Avoid purchasing PVC shower curtains (creates toxic dioxins that once in your home, release chemical gases and odors). Use shower curtains made of hemp instead (naturally resistant to mold). PVC is non recyclable and leaches chemicals that can eventually make their way back into our water systems.

smart note . Old shower curtains make useful paint drop cloths or tarps. Repurposing is part of your new mindset.

soap not dope

ALERT . some commercial products can be dangerous for people, aquifers and other living things:
"Antibacterial" soaps or cleaning products usually contain Triclosan, an antibacterial and antifungal agent. This endocrine disruptor can harm our bodies (and those of our children). It has more wide-ranging impacts after it leaves our bodies and enters water systems. Triclosan reacts with sunlight to create dioxins, a highly carcinogenic and toxic family of compounds that can react with the chlorine in our drinking water to form chloroform gas, a human carcinogen. yikes!

but is it soap?

Many commercial bar or liquid soaps are really synthetic detergent products that create suds easily and rinse leaving little residue. Pure soap is composed of alkali salts and fatty acids. There are no chemical scents or deodorants added. You may note that some soaps strip natural moisturizing oils from your skin, causing you to reach for lotions after your shower.

ah.ha

Here is an interesting thing I have learned from researching the differences between old school natural fixes and commercial products:
Chemicals are added to reduce ingredient cost and to enhance what companies sell as the "experience" of the product. These chemical illusions may also tie in with creating a false need for a secondary product to restore the effects of the first on your body.

example . soap . strips moisture . grab lotion
. shampoo . dries hair . needs conditioner

I haven't used shampoo for years and my hair is much happier (see page123).

buy . better

Dr. Bronner's and other Castile soaps are tried and true. Peppermint is my favorite soap to feel zippy clean! A 24 oz bottle will last months. (You may also purchase it in gallons, for less plastic containers to recycle.)

Locally handmade soaps are always a treat or a nice gift. Natural ingredients like coconut oil, oatmeal or honey exfoliate, clean, and moisturize your skin. An added benefit of shopping at the farmer's market is that your money stays in your community and you know what you are getting.

Avoid soaps that contain petroleum distillates or long-name chemical infusions with unspecified functions.

note: Skin is one of our largest organs. Many medications are actually applied to patches, so our skin can absorb them into our bodies. Our skin will absorb what ever it is exposed to.

Dove Soap . by Unilever . sold in 80 countries . worth $4 billion has been marketed as a mild "soap" for real beauty.

ingredient list...

sodium cocoyl isethionate (synthetic detergent)
stearic acid (hardener)
sodium tallowate (sodium salt of cow fat)
sodium isethionate (detergent/emulsifying agent)
coconut acid (the sodium salt of coconut oil)
sodium stearate (emulsifier, also used as a cheap stabilizer in plastics)
sodium dodecylbenzonesulfonate (synthetic detergent, skin irritant) sodium cocoate or sodium palm kernelate (sodium salts of coconut or palm kernel oils)
fragrance (synthetic scent, potential allergen, common skin irritant)
sodium chloride (table salt used as a thickener)
titanium dioxide (whitener, also used in house paint)
trisodium EDTA (stabilizer, used in industrial cleaning products to decrease hard water, skin irritant)
trisodium etidronate (preservative, a chemical that is used in soaps to prevent soap scum)
BHT (preservative, common skin irritant)

what's in your medicine cabinet?

We have an ongoing joke in our house in referring to old piles of over-the-counter medications (or the condiments in the back of the fridge). ***What year is it from?*** Sure, we'll give that decongestant a try, if we don't feel like driving to the store at 2 a.m. Happily, we are not dead yet from taking the outdated stuff. If you don't use it all, and most of us don't, plastic bottles of prescription medications, over-the-counter pain relievers, personal care products, and first-aid supplies get old, cluttering the cabinet.

Cull the cabinet. Remove outdated drugs and vitamins to dispose of through local collection centers. Contact your local police precinct or pharmacy to ask about medication take-back programs for proper disposal of unused or expired medications.

Never flush medications down the toilet into our water supply!

Most pharmaceuticals pass through our bodies into sewer pipes and septic systems. Wastewater treatment plants are not designed to remove these compounds. A Pennsylvania study found a dozen drugs were at levels that the researchers suspect could affect fish, bugs and other aquatic species that live in the rivers and streams where treated wastewater is often sent. Scientists in Sweden found that low-levels of an anti-anxiety drug changed the behavior of fish.

One thing is for certain . all things are connected.
What goes around will come around again.

Read the labels of products you decide to expose yourself and your family to. Here are some common ingredients it is best to

avoid:

parabens (propylparaben or butylparaben) are endocrine disruptors linked to breast cancer and reproductive issues.

formaldehyde (DMDM hydantoin or diazolidinyl urea) is used as cosmetic preservatives. Carcinogenic and causes allergic reactions.

Triclosan is in antibacterial soap and other personal care products like toothpaste. An endocrine disruptor linked to producing antibiotic-resistant bacteria. As it rinses down your drain and into waterways or our water supply, studies have shown triclosan has a negative impact on aquatic life.

Fragrance can contain virtually anything, thanks to a loophole in federal guidelines that doesn't require manufacturers to disclose the ingredients in their fragrances. The term "natural" fragrances may be more marketing than really natural.

(DEA . diethanolamine) is found in soaps, shampoos, hair conditioners, hair dye, shaving creams and skin cleansers. It can cause irritation of the eyes, nose, throat, and skin. DEA is used as a cutting fluid in oil-water emulsions, pastes, aerosols (mists) or gels. Affects the nervous, excretory, ocular, and male reproductive systems.

(TEA . trithanolamine) is found in fragrances, lotions, eyeliners, eye shadows, mascara, blushers, makeup foundations, hair care products and dyes, setting gels, shaving products, sunscreens, and skin care/cleansing products. TEA is produced by reacting with (highly toxic) ethylene oxide and ammonia (also rough stuff) creating a pH adjuster, buffering agent, surfactant (cleansing and foaming agent), masking and fragrance ingredient. Compounded exposure over time can lead to painful skin conditions or allergic reactions resulting in dry eyes, hair and skin.

(PEG . polyethylene glycol) are petroleum-based compounds commonly used as cosmetic cream bases or as softeners in pharmaceutical laxatives. PEGs may contain ethylene oxide (known human carcinogen) or 1,4-dioxane. Both degrade to persist in the environment long after being rinsed down your sink and have been classified as a developmental toxicant.

this is pretty unnerving stuff!

Phthalates ***(DBP Dibutyl Phthalates or DEP Diethyl Phthalates)*** are industrial plasticizers added to moisturize or soften skin in personal care products. A study from 1980 **(Gray & Buttersworth)** showed phthalates are reproductive and developmental toxicants in laboratory animals, particularly in males. A 2005 study reflected decreased testosterone levels among baby boys exposed to phthalates in their mother's breast milk (**Main et al. 2005**). High levels of phthalates have been found in young women of reproductive age especially based on their exposure to cosmetics, plastic wrap, nail polish and other product containing this flexibility additive. Some companies have chosen to remove phthalates from their products, but often it is a component slipped in under "fragrance."

sodium lauryl sulfate ***(or sodium laureth sulfate)*** is a surfactant and emulsifier used in many grooming, dental, or beauty care products. It is also used as an engine degreaser, floor cleaner, pesticide, or as a thickener in powdered foods and marshmallows. (crazy, right?) The Journal of the American College of Toxicology says that it has "a degenerative effect on cell membranes because of its protein denaturing properties" and that "high levels of skin penetration may occur at even low use concentration". Exposure from product has been linked to dandruff, dermatitis, canker sores and irritated tissues, eyes or skin. It bioaccumulates in our water supply and is toxic to fish and other aquatic animals. Undetected by many municipal water filters, we may be getting a second dose in the tap water we drink.

Petrolatum is mineral oil jelly used to lock in moisture, found in lip balms, skin moisturizers, and hair care products. Petroleum products can contain polycyclic aromatic hydrocarbons (PAHs), a carcinogen.

buy . better

There are lots of certified organic healthful cosmetics, dental, grooming and personal care products to choose from these days. Reward good companies with your purchase. Companies do note buying trends and there is a movement based on consumer choices for more healthful, honest products that are better for us, the earth and other living things.

or make it yourself!

See "recipes for success" pages 111-133. Easy steps and common natural ingredients to create your own in minutes.

Cleaning products for bath and kitchen

Many of the surprisingly harsh chemicals found in personal care products are also found in greater concentrations in commercial cleaning solutions.

In our "The Big Picture, recipes for success" section you can find easy substitutions for any cleaner you need, most from combining common elements already in your cupboard.

1. Toxic bleaches or abrasive concoctions to remove soap scum are harmful to breathe and worse to rinse down the drain into your septic system where they kill helpful bacteria important in breaking down waste. Consider that whatever you pour into your sink, toilet or shower mixes with reuseable graywater, which some municipalities are working to collect before it combines with wastewater. Treatment plants can only remove basic aspects of general waste. Poison is poison.

2. A second BIG plus for reassessing what cleaning agents or polishes you have collected is reducing the clutter in your cupboard and all the empty plastic bottles you have to deal with once they are empty. Many of these are not recyclable. Rinse out empty spray bottles to refill with an excellent cleanser made or tea tree essential oil, cider vinegar and water. It degreases and shines up a stove top for pennies.

3. The third win is the amount of money you will save. Polishes and cleaners are typically not a bargain and while some of the cheap ones do scour nearly anything into oblivion it's clear there is some serious toxic whatnot doing the job.

Kitchen

simple fixes

All your kitchen needs is a little green reorganization. The goals here are to reduce your stream of garbage, especially plastic. To make sure the appliances you have are used as efficiently as possible and to consider what items you bring into your kitchen. Some products we have grown to depend on are especially hurtful to the environment and can be easily replaced by more traditional means.

paper towels

Case in point: single-use, bleached paper, made from our valuable trees. Rolls come wrapped in plastic and people use loads of it. We were dismayed when a guest at one of our Airbnb lofts ordered paper towels from Amazon. They were delivered by a large truck, in a big box holding 12 individually wrapped rolls. Presumably he saved a few pennies on this purchase with free shipping? But as we all know nothing is totally free. Someone pays. With product like paper towels, pushed by heavy marketing campaigns, we pay a lot both for a throwaway product and its environmental impact.

mo . better

Your grandma might still keep a greener kitchen than many of us. My mom has a rag box and employs it. The rags come from worn out clothing, towels or linens. She cuts them into squares and stocks that box. The rags are useful to mop up a spill, clean windows, dust or do anything you might grab a paper towel for. The difference is they are reuseable and stronger so they don't rip or fray leaving lint for example all over the window you just washed. When they are dirtier than you like, toss them into a wash bag to launder. I have a hook on the inside of my sink cupboard with a dirty rag bag making it easy to collect spent rags for washing. Fresh rags might be kept accessible in a drawer or shelf, making it easy to reach for one instead of the paper towel roll.

**Save money, save trees,
each paper towel roll costs about $1, rags cost nothing.**

cling wrap

Common in most kitchens, cling wrap is a #3 plastic, single-use, petroleum product, containing diethylhexyl adipate (DEHA), a potential endocrine disruptor linked to breast cancer in women and low sperm counts in men. Don't add it to your recycle bin; #3 films like cling wrap likely have food residue that will contaminate other more valuable recyclables and can jam the machinery at MRFs.

The clingy element of plastic wrap from added chemicals and resins also makes it too complex a product to recycle. Used cling wrap either hangs around in landfills or is incinerated releasing dangerous toxins into our air. In waterways, animals eat it thinking it is food.

If you still feel enclined to use it, heads up:

plastic wrap will leach chemicals into foods, especially when it is used to cover dishes being heated in microwave ovens.

Never microwave cling wrapped food.

DEHA does get into foods, particularly fatty foods such as meats and cheeses (your sandwich or mac & cheese). Manufacturers are not required to list the chemical makeup of their plastic wrap on containers.

mo . better

Reuse fitted-top containers like those yogurt or lunch meats come in to store foods. Save screw top bottles, jars and other useful packaging in a cupboard. Wrap in wax paper or pack your lunch in a covered container. There are square sandwich boxes, sectional meal tray style containers and my favorite, stacked food tins.

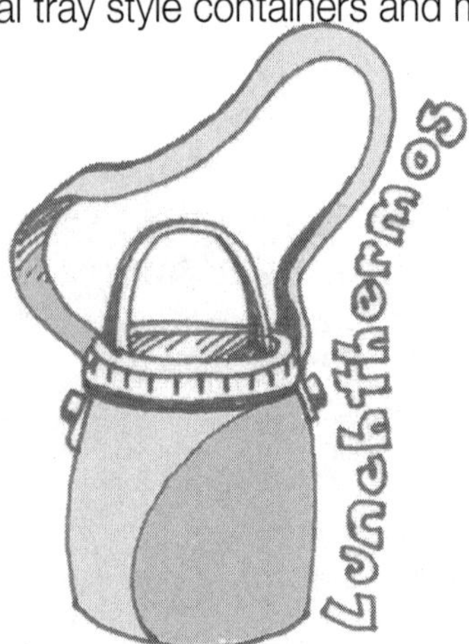

I found a plum colored one at an India baazar grocery. These are widely used in much of the world, where people enjoy leftovers and home cooked lunches. Some stack and are latched together, others like my lovely purple "Pride Food Jar" have an exterior thermos with a carry strap that the food jar fits into. Food stays hot or cold.

Look forward to lunch!

where to put recycling so it isn't a problem

If you bring it into your kitchen, you have to deal with the packaging it came in. Our heightened awareness of the importance of not wasting resources or thoughtlessly adding to our crowded landfills means recycling is necessary.

In "greening" your kitchen, make sure you have a dedicated space to collect recyclables and compost. **The zerowaste movement** *(page 55)* challenges each of us to cut down on what we throw into the garbage. Streamline recycling by providing an easy to clean transport system.

Make it . Purchase three tall plastic square waste cans with covers that pop open and close well. You can fit them in the floor area of a closet or get creative, tucking them inside a kitchen island or under a counter so they easily pull out. Single stream is not efficent, now that China declines our garbge. Separation is key (plastic, glass, metal, cardboard, paper). It is important to clean food out containers. Food waste can compromise valuable recyclables (and attract bugs).

Easy Composter

As a gardener I am horrified when I see people throw egg shells or coffee grounds into a plastic garbage bag. We keep a covered container on our counter to collect such valuable organic waste. When it is full I dump it into our yard composter. Many cities are now collecting compost to use as fertilizer or to convert to energy. It is actually cheaper to collect and convert this waste into bio-gas than to dump it in landfills, saving both our tax dollars and the environment.

The secret of getting ahead is getting started.
.... Mark Twain

Think about the end game for the items you are saving to insure they will be recyclable. For instance, glass or cardboard must not be contained in plastic bags. Containers, bottles, and cans need to have a good rinse (avoid bugs, mold, or yucky smells).

It is important to keep your materials as pure as possible. Try not to toss metal bottle caps or jar lids into the glass bin and try to refrain from stuffing the lime into your beer bottle (my bad). This mindset requires an extra step or two. If you choose to buy something that's heavily packaged it is your responsibility to return it to good use instead of to a landfill. Hopefully, companies will return to using less packaging. It seems like many items are increasingly overpackaged. Feel free to contact customer service departments with your concerns. Companies do pay attention.

we can !

Aluminum is by far the most valuable material in the recycling bin but Americans throw away more than $700 million worth of aluminum cans every year! Easy to recycle, nearly 75 percent of all aluminum produced in the U.S. is still in use today. Each can is worth $0.10! Clean off and recycle your aluminum foil too. We discard 460,000 tons of foil a year. Americans use about 100 million steel cans every day, but sadly only 71 percent are recycled. Recycling steel saves at least 75 percent of the energy it would take to create new steel from raw materials. That's enough energy to power 18 million homes.

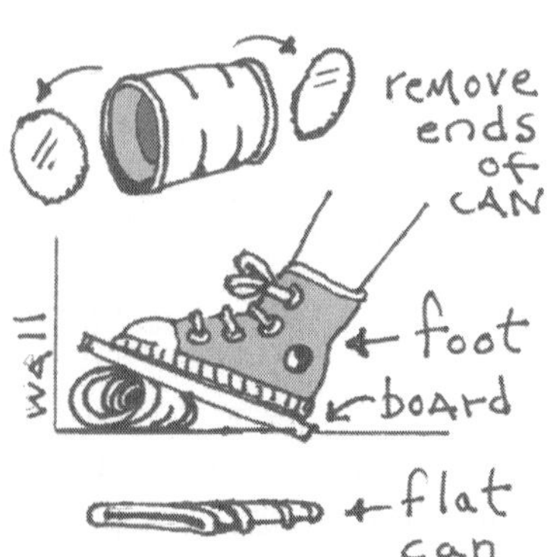

(Learn more about recycling in mining the past pages 66-73)

appliances

stove top & oven

Gas stoves win over electric stoves even though they use fossil fuels. (Electric stoves take more time to heat up & cool down.) Plan your cooking, so that you use the heat most efficiently. Scheduling a time to cook a few meals at once will save you time over the week ahead. Preheat for a shorter time or just put the food in and start cooking. Use a small toaster oven for smaller meals.

pans

I favor cast iron skillets; they last forever and cooking in them is good for you. After scrubbing with hot water, wipe with olive oil to maintain them. They cook food evenly, without sticking. You can get lucky and find them second hand. Pans with Teflon nonstick coatings will peel off into your food over time, take a pass- or opt for stainless steel.

refrigerator & freezer

Set thermostats: Fridge; 37–40 degrees, 0–5 for deep freezers,10–15 for a short-term freezer. Maintain units well, make sure door gaskets are tight, vacuum condenser coils every 6 to 12 months. Place away from heat sources or sunny windows for better efficiency.

Opt to keep leftovers in glass jars so you can easily see what is inside them. You will be able to find/grab what you want more quickly and create fewer science experiments festering behind the pickles.

If you are replacing your fridge, do your homework for an energy star approved one. The classic fridge design with the freezer on top performs 10 percent to 25 percent more efficiently than the new side-by-side models. Also, consider buying a smaller model that consumes less energy and discourages storing excess food.

Water and ice dispensers raise the unit's energy usage by 20 percent, so switch to trays, which take up less room and add more cool to the freezer. A full but not packed fridge works the most efficiently. If you live alone, add some jars of cold water to help it maintain temperature longer. If there is room in your freezer, freeze a brick-sized block of ice to help it run less and to use if you lose power during an outage.

Don't stand there with the door open or leave it open while you get something.

My dad will yell at you.

microwave

I will confess, I am not a huge fan of microwaves but they do cook quickly, using less energy than your oven or stove.When my old one failed that I had bought used about 10 years ago at the Habitat for Humanity ReStore, the new ones I looked at seemed less well made and quite pricey. After recycling the old one, I bought a new old one at the ReStore for $25. I think it used to be in a Seven Eleven, circa 1975. She isn't pretty but that girl cooks.

Only microwave in glass or microwave level ceramic. Cover with a plate or wax paper, never plastic. If you purchase a prepared meal in black microwavable plastic, never reuse it. Reheating in them will cause chemicals in the plastic to leach into your food (In fact, don't buy that prepared junk—it's loaded with salt, fats, and preservatives. Plus that black container is not recyclable).

dishwasher

Dishwashers are greener than handwashing IF you run them only when full, don't rinse your dishes beforehand or use the dry cycle. Newer models have water-miser and heat-free dryer settings.There are new very small drawer size dishwashers, but why not just wash those few dishes by hand? These seem more like another product to sell instead of a step forward. I am a fan of after-meal hand dish washing, so the food doesn't become rock hard. (Use the method next page:)

kitchen sink

Adding an inexpensive faucet aerator on your kitchen sink combines water and air to create high pressure without high water flow. Save water and rinse your dishes just as effectively. Water flows through the average faucet at a rate of five gallons per minute (or GPM), but an aerator can cut that number to around two GPM.

TRICK! Keep a big pan or container in the sink to catch water that might normally just go down the drain. For instance while you wait for the water to get hot, or if you are giving something a quick rinse. This reservoir can collect and soak small dishes or silverware used through the day for an easy wash later on.

zen cleanest greenest dish washing

Save water, time, and dream a little as you wash.

1. Use a clean washable dish rag (sponges are unsanitary)

Wash: from a small bowl filled with warm soapy water.
Rinse: in a sink of hot water with a splash of white vinegar added.
Dry: in the dish rack placed in the other sink.

2. Start with your glassware. Clean each piece and set it in the rinse sink, open sides up. Use the sprayer to rinse.

3. Put all the silverware in a larger pot that needs washing, fill with water and a little soap. Scrub the cutlery, stack silverware vertically in drainer on dish rack. Spray rinse.

4. Soap and stack dishes vertically in the dish rack. Spray rinse.

5. Do greasy pans last. To remove a hardened, burned-on crust from a pot or pan, fill it with a solution of 2 tablespoons baking soda per quart of water and bring to a boil. Turn off the heat and let the pan cool.

Air dry your dishes. This is proven to be more sanitary than drying with a towel. Speaking of linens, step away from bleached paper towel; use cheery kitchen towels instead. Instead of paper napkins use cloth napkins; keep each person's in their own napkin ring and toss in the washer after every few meals.

IDEA! A nice green gift is a series of 4 napkins each with a personalized napkin ring. Sew them yourself. Upcycle an old table cloth or material remnants.

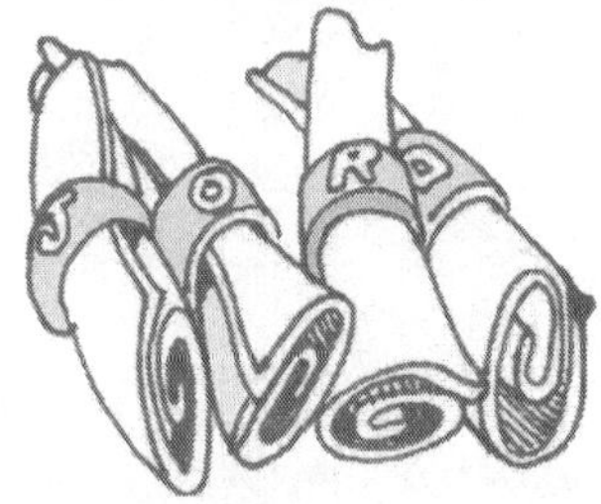

Even though paper products like napkins and paper towels are biodegradable their processing is harmful to the environment. **So remove them as a go-to and think of all the trees you'll save.**

quick fixes

Use smaller appliances whenever possible.Toaster ovens over ovens.

Plug smaller appliances (coffee maker, toaster, etc.) into a single power strip and flip the switch to "off" whenever they are not in use.

French press your coffee or use a single cup pour through if you don't need a whole big pot. Cold brew it overnight for a mellow cup. It is especially lovely iced. Steep grounds in water like tea.

Crockpot a big soup. Freeze half for future meals. Label/date clearly.

Cover pots with lids to cook faster.

Keep your market bags in your car to use shopping. Practice using them and not just at the grocery: anywhere someone wants to hand you a plastic bag.

Buy bulk non-perishable items to cut down on packaging.

Buy local fresh food for just a week for easier storage and less waste.

Switch to biodegradable garbage bags so contents (and bag) decompose in landfills.

Reuse netted onion bags for packaging grocery produce instead of taking their filmsy plastic bags. Cut off the top and thread string through the netting for a drawstring. Use these to transport loose veggies home from the grocery store.

what's cooking

BIG box stores have been training us to buy BIG amounts of food that is double packaged, full of preservatives and hard to store. We have purchased BIGGER refrigerators, more cupboards and extra freezers. The sell is that buying MORE will save you money, but sometimes folks end up throwing out portions they can't eat in time.

When I lived in Italy, my favorite part of the day was going to the market and dreaming what I might cook that evening with the fresh produce in my basket. Farmer's markets here offer a similar experience, plus you are buying local, supporting small farms, and eating healthy.

buy less, store less, eat fresh, be happy

When you do fire up the stove or oven, cook several things at once making a big batch to freeze in small portions. I like to make enough for a few lunches, along with extra servings wrapped in wax paper to drop off a meal at an elderly neighbor's or leave a treat for a busy friend. Soups are wonderful for combining leftovers into a new yummy meal. Clearly label & date stuff you freeze. I've been known to harbor lots of mystery soup of unknown origins.

Not Cool . Some veggies benefit from staying out of the refrigerator: tomatoes, cucumbers, eggplants, peppers, onions, shallots, garlic, potatoes, summer fruits, winter squash, melons, avocados and basil. Farm fresh eggs can stay out of the fridge until you wash them.

Kitchens

People like to upgrade their kitchens with the latest energy star rated appliances, trends in cupboard decor and counter tops. If you do decide to go this route donate any castoffs to your local Habitat ReStore, or if things don't work anymore see if there is an appliance recycling program in your area. In redoing your kitchen, first visit a ReStore to see if you might find some lovely cupboards someone else has torn out. (Many new cupboards have a high formaldehyde content, so try to purchase real wood or wheatboard.) In renovating the upstairs of our studio and store building, our family honed our carpentry skills to create four elegant Airbnb lofts, with kitchen cupboards and most of the furniture furnished through our area ReStore. This not only saved us loads of money, but amused us through the scavenger hunt challenge of puzzling together each space.

Habitat ReStores are a win/win. You support Habitat in their efforts to build community housing while supplying the ReStore, instead of the landfill, with your castoffs. Plus, it is exciting to find useful treasures at low prices. Besides kitchen cupboards and furniture, they stock construction materials, working appliances, windows, doors, and light fixtures. There are also used tools and paint, including full gallons and mis-mixed colors donated directly from paint stores.

. delightful oops paint .

This is a great way to get the color you want in any room. If you find a light color, a paint store will usually mix color it for a small fee. Or be adventurous and mix it yourself. Experiment, noting the portion amounts you add in small paper cups until you get the color you like. Paint a section of the wall. Over the next 24 hours see how it looks in sunlight and room light. Mix up a full batch and jazz up that room! You have saved money and used perfectly good paint that is otherwise a hazard to dispose of.

Paint is my favorite solution for a makeover that spruces everything up by using what you have.

lighting

Natural light is always best. Set up your home to take advantage of it. Artifical lighting has issues, so use it only when needed.

LEDs offer a bright white light and are safer than compact fluorescents, but CFLs are more energy efficient. There has been a big effort to move light bulb usage from incandescent bulbs to these. But......

...disposal or breakage releases mercury.

Excerpts from the EPA website:

If you have a broken CFL . BEFORE CLEANUP

Have people and pets leave the room. Air out the room for 5-10 minutes by opening a window or door to the outdoor environment. Shut off the central forced air heating/air-conditioning system. Assemble stiff paper or cardboard, sticky tape, damp paper towels or disposable wet wipes (for hard surfaces). During cleanup Do NOT VACUUM. Scoop up glass fragments and powder using stiff paper. Use sticky tape to pick up any remaining small glass fragments and powder. Seal the used materials in a glass jar or plastic bag.

To dispose of call your local poison control center at 1-800-222-1222 or a waste management company ....yikes!

Lighting containing mercury

Fluorescent bulbs:

Linear, U-tube and

circline fluorescent tubes

Bug zappers

Tanning bulbs

Black lights

Germicidal bulbs

High output bulbs

Cold-cathode fluorescent bulbs

High intensity discharge bulbs:

Metal halide

Ceramic metal halide

High pressure sodium

mercury vapor

Mercury short-arc bulbs

Neon bulbs

Opt for green energy

Over half of the electricity in the U.S. is produced by burning coal and natural gas, which releases carbon dioxide and other pollutants into the atmosphere. To reduce your carbon footprint, consider purchasing green power for your home instead.

Renewable energy is collected from resources that naturally replenish on a human time scale, such as solar, wind, Hydropower (flowing water, tides, waves), biofuels (plants, algae), geothermal heat, and Biomass (recently-living natural materials like wood waste, sawdust and combustible agricultural wastes). Fossil fuels are a finite resource taking millions of years to develop and release climate warming carbon into the atmosphere.

Most utility companies have a green energy option that allows you to purchase your electricity from a renewable energy source, such as wind or solar, rather than power generated by fossil fuels. Advances in renewable energy technologies have lowered their cost. Green energy can now replace fossil fuels in all major areas of use including vehicle fuel, electricity, water, and space heating. To find out if green electricity is available in your area, go to the state by state map at Buying Green Power on the U.S. Department of Energy website.

If you wish to opt for residential solar panels
be sure to research the company, read reviews and make sure that you fully understand future maintenance contracts.
Choose a reputable solar company to work with.

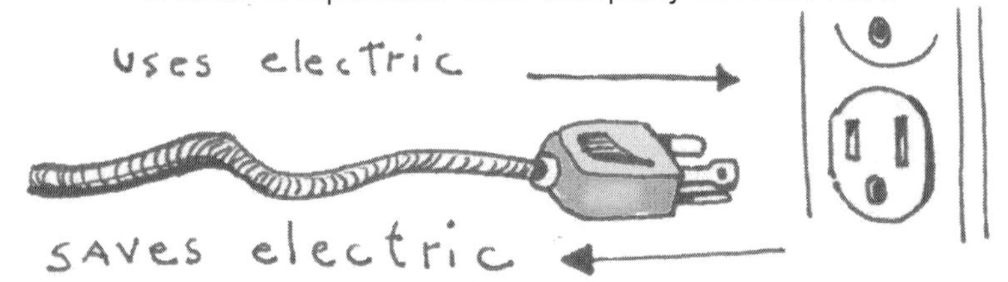

Get into the habit of unplugging an appliance right after using it.
Anything that plugs in uses electricity
even when the power isn't on.
40% of the energy that home electronics use
is when they are not even on. OUCH!

bedrooms and throughout your home

Moderate your home. Keep it cooler in winter. Add more blankets to the beds, wear slippers and toss on a robe when you wake up. (Cooler home temperatures contribute to better health.)

It's nice to be surrounded by your fans!
Run your ceiling fans clockwise on low during winter to push the warm air down and counterclockwise on high in summer to feel cooler. Open your windows in summer and direct fans to blow toward you. Use AC sparingly. (Our bodies acclimate to temperature. You will find you are far more comfortable in summer the less time you spend indoors with AC). Use thicker curtains on windows in winter to help trap heat inside. Utilize blinds in summer to deflect heat back out.

get small . If there are rooms you rarely use, close them off during parts of the year requiring extra energy to make them comfortable. We hunker down in about three cozy rooms during winter. If guests stay, it's easy to fire up extra rooms.

thermostat . A programmable thermostat allows you to turn down or off the heating or A/C when you're not home.

Rugs . Area rugs help to insulate the floors of your home.

Sweep . Hand broom floors and stairs. You get the dirt and some exercise. Leaving your outside shoes by the door keeps dirt and toxins from being tracked in. Perk! Less vacuuming.

house plants . Improve inside air quality and your disposition. Pour gray water or old coffee in plants rather than down the sink.

laundry

The average household does almost 400 loads of laundry each year, consuming about 13,500 gallons of water.

Your clothes will last longer if you wash them less. Wash things when they are dirty, not after wearing once.

Try to wash ONLY full loads for shorter cycles using cold water.

Use a small amount of concentrated phosphate-free, plant-based biodegradable soap. Conventional laundry soaps can cause algal blooms that negatively affect ecosystems and marine life.
Or make your own laundry soap! ***Blend in a food processor :*** 1 cubed bar of castile soap, 14 ounces borax & washing soda. Add 10 drops of Teatree essential oil. Use 2 tablespoons per load.

Replace fabric softener with a cup of white vinegar added to the washer during the rinse cycle. Vinegar naturally balances the pH of soap, leaving your clothes soft and free of chemical residue.

Use a clothesline and fresh air to dry your clothing to save energy. You can puff any stiff laundry with a brief spin in the dryer.

Use a drying rack placed in your shower or tub to dry delicate items or sweaters. They will look nicer and last longer.

Smooth out and hang shirts or items that wrinkle when they are damp to avoid ironing them. Damp fold cloth napkins and pillow cases for crisp edges.

Avoid purchasing things that need dry cleaning. Instead of using bleach, use sunlight. Released into waterways, chlorine bleach can contaminate drinking water creating organochlorines. These suspected carcinogens are reproductive, neurological, and immune-system toxins known to cause developmental disorders. Once introduced into the environment it can take years, or even decades, for them to break down to less damaging forms.

the office

Apply what you do at home to where you work, plus these ideas...

Minimize your use of paper. Print on both sides. Or reuse by cutting up into smaller pieces for messages or lists. Finally collect to recycle. Purchase unbleached recycled paper stock.

Use recycled toner cartridges when printing is necessary. Recycle used toner cartridges and batteries thru Staples or Office Depot.

Donate old office equipment and computers to training centers, or Habitat, or recycle them.

Utilize file-sharing websites like DropBox or Google Docs instead of printing or maintaining paper files. Use your company's network, zip drives, external harddrives or the cloud to store documents that you'd otherwise print and file.

Plug all of your electronics into power strips that you can easily turn off at the end of the day. Turn off lights unless they are in use.

Fill the kitchen area with ceramic plates, bowls and mugs, glasses and metal silverware. No single-use products or plastic.

Ban bottled water. Use tap or a water cooler with washable glasses.

Set up a recycling station. Save and reuse packaging materials or donate to a neighbor business that ships, rather than tossing them.

Stop printed catalogues, flyers, and junk mail coming to your address. (See **enough already** page 76)

Employ outdoor temperatures, keep the thermostat moderate. Open windows or wear sweaters. Utilize daylight where possible.

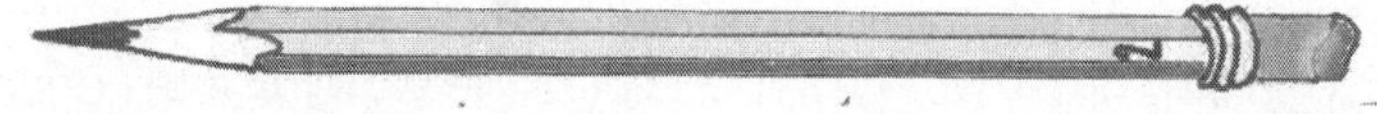

Use refillable pens and pencils, or good old wooden pencils.

lawn & garden

For centuries, most buildings in dry climates have had rainwater collection systems feeding cisterns in their basements. This smart practice would reduce water usage, issues with rainwater runoff and during a crisis, alleviate water shortages.

(see: **a perfect world** page 104)

install gutters & rain barrels

Collect rainwater and use it to water your garden.
Yes, hand carrying watering cans is work or is it...exercise! You can starve the weeds, just nourish the plants you pour water on and enjoy your garden instead of going to the gym!

Like many residents of low-lying areas near the ocean, we have issues with salt water encroachment in our water table and well. For years, I was dumbfounded as to why my garden looked so sickly after all my hard work and careful watering with the hose. My friend, Randy, who ran a plant nursery for many years, frowned at the brown tinged leaves on my tomato plants. "Your groundwater is too salty; if you use it, it will kill your plants."

So last summer we installed gutters that emptied into jolly blue garbage cans ($10) with lids, so as each one filled we covered it. This keeps the mosquitoes from breeding within. I asked for a lovely aluminum cow feeding trough for my birthday. It captures 200 gallons of water. It has a wood top with a hinged panel allowing me to dip my watering cans in. I spot water the garden, starving the weeds, but feeding my veggies. This exercise has the added benefit of a daily mini-workout for my core and arms.

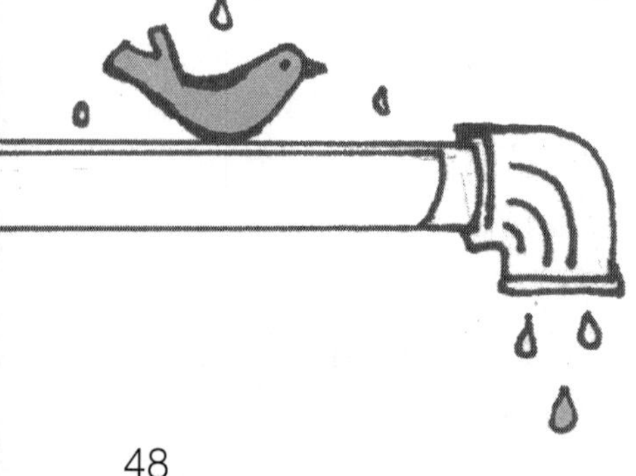

We even rig the gutters to fill our swimming pool each summer. The rainwater requires far fewer chemicals to keep the pool clean and balanced. Now every time it rains, I am doubly excited. When all the water containers are brimming, I feel rich.

tree hugger

Trees are the salve for our earth. The over-abundant carbon dioxide produced by our dependence on fossil fuels is utilized by trees during photosynthesis cleaning the air and restoring oxygen to our atmosphere. Trees are the lungs of our planet, as well as an important renewable resource. The more of them the better.

If you have a yard, tuck in a new tree or shrub anywhere you can. There is something magical about bringing in a pile of fresh figs and pears from the front yard! Flowering trees are so uplifting. Plant trees to shade your home in summer. In winter, the sun will warm your home again shining through the bare branches. Fruit or nut trees will also offer a harvest. When we buy a new house the first thing we do is plant some trees. Our current house was down on its luck, empty for years, waiting patiently in the middle of a field at the end of a dirt road. There were a bunch of sad trees on sale outside of Big Lots. Some cottonwoods, a maple, and three crabapple trees.

I spent $60 and lugged them home to plant. Twenty years later, the cottonwoods are easily 80 feet tall, the crabapples treat us to a flurry of flowers right around Easter every year, and our dirt road is an elegantly shaded—well, um, dirt road.

Dig a hole twice the size of the root ball, loosen the soil around and under where the tree will go. Drag over a bucket or two of water to fill the hole. As it soaks in, unwrap the root ball and loosen the roots if they are thickly bunched. Put some compost in the hole and set in the tree so the top of the root ball is just lower than the ground. Replace the dirt, adding more compost and water. Mound dirt up to trap water, fill with mulch and voila! Baby your new little tree—make sure it has enough water, especially that first year.

good 4u

Gardening is more of a fruitful yoga than work. You stretch and pull, focused on the work before you, pleasantly absent from any worries. As a lazy gardener, I bury cardboard under my mulch to deter weeds. Most of my garden is made up of unfussy indigenous plants that enjoy coming back the following year as volunteers. A composter makes short work of our biodegradables (leave animal products out—just kitchen scraps, leaves, and plant material).

If you don't have space or time for a garden, set a few large pots somewhere sunny. Growing things is good for you. Herbs can tuck in anywhere or live in containers inside or out. Clipping some fresh herbs to add to a dish makes the meal extra special. A vase full of fresh flowers you have grown is a lovely antidepressant. Heaven is a handful of ripe rasberries picked on a sunny morning or crushing mint leaves into your iced tea. Both fuss-free plants grow prolifically whether you invite them to or not.

IDEA! Gather your gardening pals. Have everyone bring their extra seedlings, volunteer veggies, culls from crowded bulb gardens, or whatever they wish to share. Give and get! This is a great way to save money and make your garden deeply lovely. You can enjoy that magnificent lily and recall who you received it from.

mow less

Limit your lawn area. If you have a big yard, consider turning part of it into a wild flower area (mowed one time a year) or better yet plant trees or shubbery for wildlife to enjoy. Leave the lawn clippings to mulch the yard or if you choose to rake them up, reuse them around the base of trees. Sculpt portions of your yard into outdoor rooms by adding patio areas, fishponds, walks, and bulb gardens. Raised beds are easy to tend and look great. Cover the pathways in between them with mulch to stiffle weeds. We have built a garden island in the middle of our yard with our house in the center. The grass is never invited in.

If your yard is small, consider getting an old fashioned push mower to tighten your abs while you manicure your little lawn. People energy is good for the environment and for the people too.

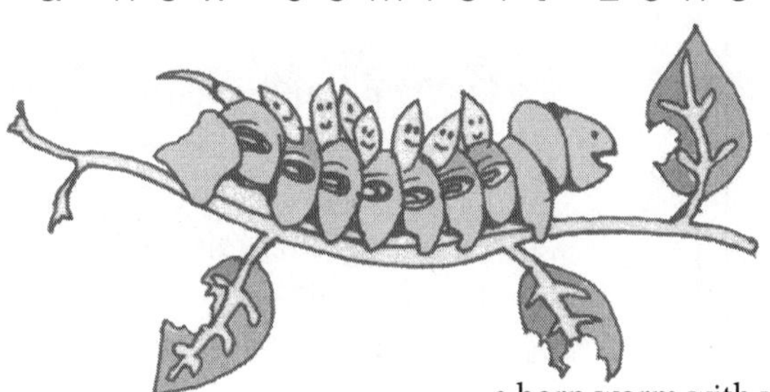

a horn worm with wasp larvae

love bugs

Most bugs are actually good for your garden, some will handily take care of bad bugs. Tomato hornworms chow down on your tomato plants. If your plant suddenly looks like it was dinner for someone, look sharp for those well-camouflaged bandits. But inspect that varmint before you dispatch it. If its back is covered with lots of white larvae, clap your hands and look smug. Those are parasitic wasps to the rescue, who will have that hornworm for breakfast. Sure, caterpillars eat leaves, but butterflies are such a treat! (Remember caterpillars feed songbirds too.) There are many more pollinators than just our beloved honey bees. It is a good thing if your garden is buzzing. So don't be bugged, just bee.

mother the earth

Go organic. Pesticides will upset the natural balance of your garden and sully the soil. For centuries gardeners have tended their crops using a variety of practices to enhance their harvest, while lessening the work of it. Vinegar or very hot water poured on weeds kills them. Covering an area with a tarp or cardboard and mulch kills weeds in a few days. Set your garden beds up so there is little space for weeds, heavily mulched in between plants, to both retain water and nourish them. I am a big fan of straw, especially in the vegetable garden. Pine straw or homemade leaf/dirt/organic waste mulches are always a plus for the soil.

A loop hoe easily keeps plant beds tidy. Spray a mixture of chili powder, crushed garlic, dish soap and water to keep bugs off your vegetable plants. If beetles are feasting away pluck each one up and imprison it in a jar of water. No one ever said the earth was just for us.

Coexisting is far more interesting.

proactive convenience

Convenience is marketed to us, but who does it really benefit? Let's explore how to easily retool by taking proactive steps to define what is truly convenient. How to live and eat well today, as well as in the future.

If we are honest, I believe that deep down each of us is aware that we have been taking more than our fair share of everything, crowding out other living things for our own comfort and personal consumption. Our dependence on damaging chemicals, plastics and fossil fuels has polluted air, water, and land. We have been the thoughtless bully on the playground.

Now, the very things that have sustained us have grown angry and unpredictable. Normal weather patterns swing to extremes producing record droughts and drenching storms. Animal species are facing extinction from over hunting, food scarcity, and habitat loss.
The oceans rise as the ice caps melt. The sun's rays have become harmful. We have begun to reap what we have sown especially during this last century.

yet-

The human population continues
to balloon without conscience.
Some of us are overly fat, indulged and happy
while others struggle to live.

A new mindset is in order.
It is time to readjust our daily practices
to greatly reduce the footprint we leave behind.
This mission begins based on taking the long view
instead of looking for short-term personal gain.

Let us explore . what is enough.
Moderation is a personal responsibility.
The earth is for all.

We are what we
repeatedly do.
Excellence, then,
is not an act, but a habit.
. . . . Aristotle

Zero Waste is a philosophy that encourages the redesign of resource life cycles so that all products are reused. The goal is for no trash to be sent to landfills, incinerators, or into the ocean. Nature's cycles that replenish as part of their circular process can be the basis for the most efficient inventions.

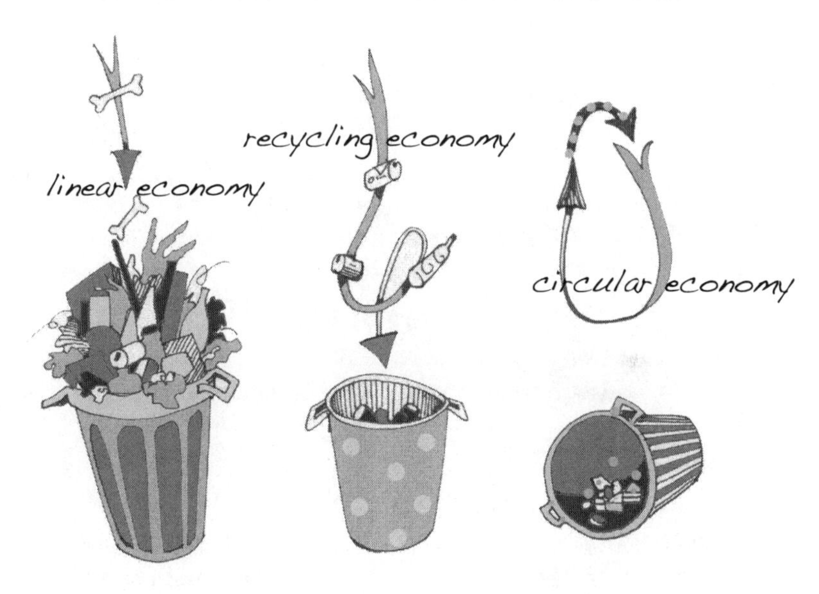

To play this game it helps to think proactively about what you want, if you really need it, and what you will do with what is left of it post use. Our current habits of convenience involve making a big stinking mess that someone else will have to deal with later.

Instead of convenience this is avoidance.

blinded by marketing

We've been sold a package of NOT GOODS

Plastic Razors are an example

of absurd consumption.

They will never return to soil and are used only a few times before being discarded. People purchase plastic razors in plastic bags of 30 to toss each after a few shaves.

2 billion razors are thrown away each year

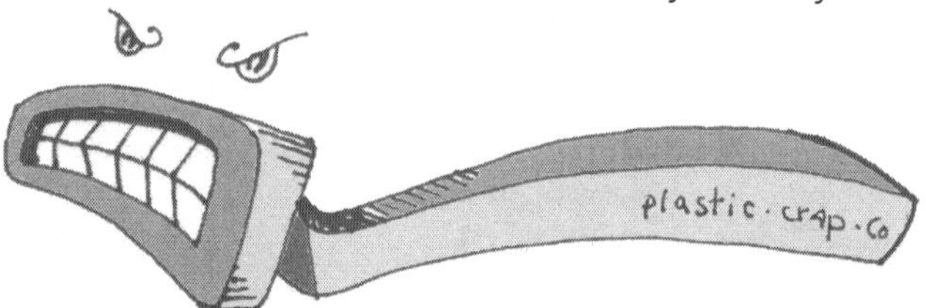

Before 1901, when American inventor King Camp Gillette patented a new variation of safety razor with disposable blades, men shaved with straight razors. Gillette realized a profit could be made as each blade wore out and was discarded. In the 1970s, Marcel Bich, the founder of the Bic company in France, decided to begin creating a completely disposable razor. The Bic razor was sold as an affordable product that simplified daily actions.

Really?

Shaving is shaving. Disposable items mean profits for whichever company is hawking them. Do we want to support them in creating a product that is environmentally irresponsible? Bic argues that

"Our razors are too small and lightweight to meet recycling criteria, as recycling is only justifiable for products at end of life when they offer significant potential in both weight and volume, or are easy to disassemble."

This justifies what exactly? Money in their pocket.

Green Gift! A new metal safety razor will last forever (cost: $30-$80, replacement blades: $1 last 1–2 months). Bags of disposable razors cost $7–10 dollars each time you purchase them (your money to the landfill). Electric razors are another option.

The massive plastic islands of trash floating in the oceans were built by us piece by piece.

Advertisers have done a great job convincing us to opt for their prepackaged meals, single-use products and extraneous fluff. It's interesting to note that quite often the less packaging the better an item is for you. For example, fruit comes in its own wrapper. It feeds the earth along with you, spreading seeds while fertilizing the soil.

If you buy bulk unpackaged goods you'll save money and recycling chores later by bringing your own containers.

the raw food movement

Some people believe that heating food destroys nutrients and natural enzymes that boost digestion and fight chronic disease. A raw food diet consists of uncooked, unprocessed, mostly organic foods. Staples: raw fruits, vegetables, nuts, seeds, and sprouted grains. Some will eat unpasteurized dairy foods, raw eggs, meat, and fish. The food can be cold or even a little bit warm, as long as it doesn't go above 118 degrees.

Eating lots of veggies and fruits is a more natural way to eat and contains less sodium. They are also very good for you, helping to control blood pressure, lowering your chance of stroke, heart failure, osteoporosis, stomach cancer, kidney disease and can even prevent type 2 diabetes. At the least, upping your raw food intake will help you feel better, take off weight, and generally steer you toward a healthy lifestyle.

Let things taste
the way they are.

.... Alice Waters
chef

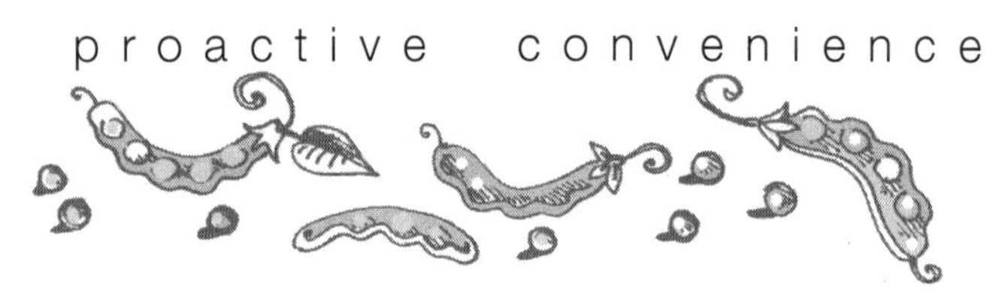

If you stock your kitchen well with fresh produce, basic grains, legumes, and spices, delicious dishes can be mobilized nearly as quickly as preprocessed meals. Studies have shown that rushing to eat any food is far less healthful than the fufilment one receives in preparing and eating food calmly. The return to slow food is a reaction by many of us to growing up in a fast food nation.

the slow food movement

was started in Italy by Carlo Petrini in 1986. The movement is now international, envisioning a world in which all people can access and enjoy food that is good for them, good for those who grow it, and good for the planet.

Slow Food is based on three interconnected principles:

GOOD:

quality, flavorsome and healthy food

CLEAN:

earth friendly production that does not harm the environment

FAIR:

accessible prices for consumers
along with fair conditions and pay for producers

** Food for Change is the first international communication campaign that links global warming with food production and consumption, encouraging a more reasoned approach.Their many projects include ways to support family farming, food/environmental education, and biodiversity protection.

New food movements signal a return to local, decentralized natural food practices.
How we used to live.

One of the biggest culprits in climate change is the industrial food production system.

Fully one fifth of greenhouse gas emissions are produced by intensive livestock farms, the widespread use of chemical agents on crops, produce grown year-round regardless of its natural season, and the lengthy supply chains that fly food across the planet to our supermarket shelves.

A visit to a milking facility in upstate NY where 4,000 cows were housed in massive cement barns designed to pull rivers of their manure into storage "lakes" deeply unnerved me.

Our hearts must harden to sanction a lifetime for any animal where grass is never felt underfoot and the warmth of the sun is always just out of reach.

Unfortunately traditional agriculture, particularly small-scale farming, is also the first victim of climate change. Farmers have to deal with devastating droughts or overly wet periods. In arid climates they must make longer and longer journeys to find water for their animals. Rising sea levels threaten the survival of fishing communities, while the acidification of the oceans is making them hostile to life.

Earth is already experiencing a rise in biodiversity loss and unstoppable desertification as precious forests are cut down for wood or to make room for monocultures, grazing, or more farmland.

Each one of us can make a difference in our daily shopping and lifestyle choices. The combined economic effect of many can push governments and the international community to legislate the positive changes that our planet so desperately needs.

Animals are my friends...
and I don't eat my friends.

....George Bernard Shaw

eat well

**Maybe we can redefine what it means to eat well.
It used to mean having the resources to dine high on the hog, with an abundance of treats to choose from. This concept of food entitlement became a "more is better" license to market oversized portions of rich processed food and bucket-sized drinks full of sugar.
The effects of this are visible daily in our overweight population and diabetes epidemic.**

There is a strong movement to exercise and eat well, but it is not a mindset that extends to those with less privilege. Sugar and salt are clearly addictive. Both are deepy entrenched in the American diet and economy. So once again, it is the individual that must choose.

What is best for my health and the health of my family?
Avoid processed packaged foods or commercially prepared meals. Eat well by serving fresh, locally sourced home-cooked meals in reasonable portions. You will feel better, save money and help the environment.

the side effects of convenience

Factory farms that stock grocery stores are designed to produce mass amounts of food in the cheapest way. Animals are overcrowded under inhumane conditions—their numbers producing a density of waste that harms the environment. Big farms love single-crop applications boosted by chemical fertilizers and pesticides that strip the soil of natural nutrients. Soil degradation through tilling and chemical use contributes to erosion and barren stretches, where plants struggle as moisture is wicked out leading to desertification.

Pesticides kill bugs (both helpful and others), birds, and small animals. These chemicals can also harm larger animals (like us) over time.
Fertilizers wash into waterways causing massive algal blooms that rob oxygen from the water causing hypoxia, killing fish and other living organisms. Single crops crush area biodiversity while demanding huge amounts of water, much of which is irreplaceable fossil water pumped from deep underground.

maybe a better term is to eat mindfully

When possible, buy in-season local produce and organic meat.

An EWG (Environmental Working Group) study ranked fruits and vegetables by the amount of pesticides found on them.

wash it well

Highest Pesticide ranking from worst to better:

Peaches, apples, sweet bell peppers, celery, nectarines, strawberries, cherries, lettuce, imported grapes, pears, spinach and potatoes.
I have offen stuffed down nearly all of these without a rinse, especially peaches and strawberries. If organic costs a few cents more, at least you can be clear about exactly what you are ingesting. You can get away with grocery broccoli, asparagus, onions, avocados, pineapples, mangos, frozen peas or sweetcorn, kiwis, bananas, and cabbage.
They fare lower in the ranking.

We are what we eat eats.

.... Michael Pollan's
The Omnivore's Dilemma

sourcing local meat & dairy

Many farmer's markets now include local dairy farmers along with meat producers. Some dairies offer refillable glass milk bottles, along with, artisan cheeses, yogurts, and ice cream. Local meat producers sometimes offer CSA type subscription programs. For instance, you can purchase half a cow pre-processing; your half will be butchered and packaged in butcher's paper. Check Eatwild's Directory of Farms (eatwild.com). They maintain a growing list of pasture-based farms for grass-fed meat and dairy products in the United States and Canada. Products include: beef, pork, lamb, veal, goat, elk, venison, yak, chickens, ducks, rabbits, turkeys, eggs, milk, cheeses, wild-caught salmon, and more.

This is a simple return to village markets and small agriculture. Shopping fresh makes for inspired meals.

best fishes

fish and seafood . Fish stocks, historically a staple for many cultures, are extremely low. I love all sushi, but now I limit fish content fare and stick to california rolls. Only 10% of tuna are left after 70 years of industrial fishing. The Blue Ocean Institute (safinacenter.org/programs/sustainable-seafood-program) monitors 90 seafood species for current sustainability levels and offers a safe eating rating. Farmed fish like salmon or tilapia can contain PCBs, dioxins and pesticides. It is a good idea to know the backstory on your fish.

Eat without eco-guilt

Atlantic Mackerel
(purse seine from Canada and the US)

Wild-caught Salmon, fresh or canned and Freshwater Coho Salmon. (farmed in tank systems, from the US) **AVOID:** Atlantic Salmon or imported salmon raised in open-net ocean pens, which live in a wash of waste, chemicals, antibiotics, diseases, sea lice, and parasites.
Farmed salmon can escape into the sea infecting other fish.

Atlantic Herring & Pacific Sardines (wild-caught) . low in mercury and high in heart-bolstering omega-3s. Very tasty in pasta or atop a salad.

American and Canadian Pacific spot prawns . Shrimp is one of the top seafoods we consume but is nearly 90 percent imported. Fine-mesh nets kill more pounds of “bycatch” than shrimp. The wasted bycatch is tossed overboard. Farmed shrimp is sometimes sold as wild or Gulf shrimp. This mislabeling means you may be unknowingly eating antibiotics, fungicides, and other harmful chemicals used in farming shrimp.

Albacore Tuna . (troll or pole-caught, from the US or British Columbia)

Sablefish/Black Cod . (from Alaska and Canadian Pacific)

Seabass . caught by small day-boat American fishermen and marketed directly to consumers through community-supported fisheries (CFS).

farm-raised Catfish, Oysters, mussels or clams . small and sustainable
Eat low on the food chain, smaller seafood contains less mercury.

back to the future: grains & plant proteins

Grains are the powerhouse of our sustenance and they are far less resource-intensive to produce than animal product foods.

0.51 liters of water to produce 1 calorie of grain, 1 calorie of beef requires 10.19 liters of water

Grains can store longer, are easy to transport and versatile in cooking. Whole grains improve the health and diversity of your gut microbiome, lowering your risk of chronic disease, cancer, obesity, heart disease, and stroke. Many small grain producers cultivate using no-till, or "direct seed," practices. This reduces erosion while building up the soil's organic matter eliminating the need for chemical fertilizers. Growing oats, rye, wheat, or triticale during a second growing season replenishes the soil, yields a crop, and traps more carbon matter. The results of a three-crop rotation required 86 percent less mineral nitrogen fertilizer, reducing nitrous oxide emissions, and 96 percent less herbicide use. Grains are pure goodness.

You can find an astounding variety of grains in stores or online! Grains and where to purchase them . wholegrainscouncil.org

Amaranth (Amaranth Flour, Amaranth Grain)
Barley (Barley Flour, Barley Grits/Meal, Hull-less Barley)
Bulgur (Bulgur, Golden Bulgur)
Buckwheat (Buckwheat Groats, Buckwheat Flour, Kasha, Creamy Buckwheat Cereal)
Whole Grain Einkorn
Kamut® Wheat (Kamut Berries, Kamut Flour, Kamut Cereal)
Flax
Whole grain or cracked freekeh
Millet (Millet Flour, Millet Grits/Meal, Hulled Millet)
Oats (Whole Groats, Oat Flour, Steel-Cut Oats, Quick-Cooking Steel Cut Oats, Scottish Oatmeal)
Wild or Brown Rice (Long Grain, Short Grain, Basmati Long Grain, Brown Rice Flour, Sweet Brown)
Quinoa (Red Quinoa, White Quinoa, Tri-color Quinoa, Quinoa Flour)
Rye (Rolled Rye Flakes)
Sorghum (Sorghum Flour, Sorghum Grain)
Spelt (Spelt Flour, Spelt Berries, Rolled Spelt Flakes)
Teff (Teff Flour, Whole Grain Teff)
Triticale (Rolled Triticale Flakes)
Wheat (Hard Red Wheat Berries, Soft White Wheat Berries, Cracked Wheat, Rolled Flakes, Flours)
Rye (Rye Berries, Cracked Rye, Dark Rye Flour, Pumpernickel Dark Rye)
Grain Blends (Whole Grain Medley)
Cereals (Whole Wheat Farina Cereal, Brown Rice Farina Hot Cereal, 5 Grain Rolled Cereal)

good 4 u

Whether you are vegan, vegetarian, a locovore or meat lover, food not only sustains us it can bring us together. Take time to sit down to eat with your family, friends or coworkers. When I studied abroad in Florence, Italy, during college, I was impressed to note that restaurants opened for dinner after 4pm, allowing the entire staff to sit down and eat together before any customers arrived. Differences were ironed out, comaraderie was enjoyed, and no one worked hungry like many of us who have waited tables in lean times. It is always good for chefs to taste their own cooking and for the wait staff to be familar with what they are serving.

The shared meal elevates eating from a mechanical process of fueling the body to a ritual of family and community, from the mere animal biology to an act of culture.

.... Michael Pollan

Durning my time abroad, I loved the ritual of collecting fresh ingredients at the market square, deciding a menu, and creating a meal together with friends. The inviting fragrances and colors transported me to a place of joyful anticipation the minute the garlic hit the olive oil. The midday meal was the big one. Everything in the city stopped as kitchens filled with laughter and stories. I nearly didn't come home to the United States.

When we have big dinners the kitchen becomes the center of creativity. The psyche of food is also part of its nutrition. Making a meal is putting your love into a dish you will share. It is a treat to see a proud cook set her pie down like it is a golden gift.

One taste and clearly it is.

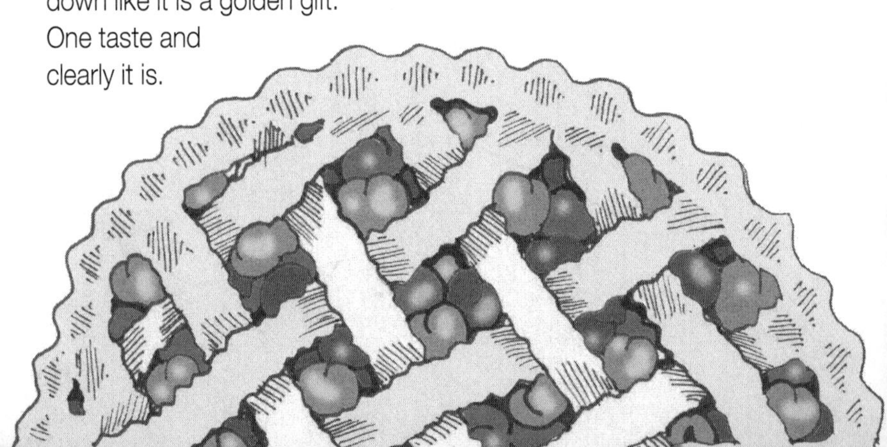

come together and eat

Why not have **"bring whatcha got"** dinners with your friends or neighbors once a week or month? These brighten up the work week and everyone eats well. I just send a group text out saying something like: "I have vegetarian chili" and happy replies roll in immediately with what others will bring including some lovely wines or beers we all must try.

eating out . bring it

Happily, the restaurant world has finally gotten the memo to move beyond polystyrene takeout boxes and plastic straws. The United Kingdom has completely outlawed single-use plastics and many states here in the US are moving in the same direction. The Queen of England has banished straws. The city of New York nixed foam cups. Some establishments may lag behind. We can encourage them to return to paper takeout containers, but one step better is to bring your own covered drink cup or food container with you. Takeout cups and containers are a pricey purchase for restaurants as well as for the environment. New biodegradable plastics need special conditions to break down. Being buried at a landfill will not work.

Tuck a cloth napkin with metal utensils into your briefcase and leave the plastic utensils alone. If you do need to take some, wash and reuse them in your lunch kit.

Travel with a big stainless steel insulated covered cup. Your drink stays cold or hot. Some cafes will give you a discount for bringing your own container. Keep it filled with water as an alternative to the horrible plastic water bottle craze. If you must have a straw, choose a paper or stainless steel reuseable one.

mo better

Why not spend a few dollars at Goodwill or another secondhand store and buy vintage ware instead of disposable plates, silverware, or cups. I found 6 different wine glasses at fifty cents each to offer to fellow members of our Sierra Club steering committee. Each person got to choose one (along with a beverage to fill it). These made our meetings extra fun and productive. Many thrift stores' profits benefit charities like homeless shelters or humane societies, a far better way to spend your money than on more plastic red cups.

There is so much stuff in the world right now,
if you need something you can probably find it
out there in the world of whatnot, giving it a second life.
When you are done with it, pass it on.

Proactive convenience means pre-thinking how to make aspects of your life more efficient and pleasant.

When you are out and about, combine errands. If you are going to a store where a friend or someone who has trouble getting out might need something why not pick it up for them while you are there? Others might return the favor for you on a busy day.

Coming together to fight climate change
can have the added benefit of building community and reducing
the feeling of isolation some people feel.

Proactive convenience means working together to benefit the greater community.

Keep all your bags in a sturdy market basket.

(Great for carrying all the heavy things like milk, wine or large items.) Help the cashier by packing your own bags. Sometimes cashiers are thrown off their rhythm by being presented with something other than reaching for a plastic bag. This moves the line along and you can put like items together for an easier time putting stuff away when you get home.

Mining the Past

In confronting an uncertain future, we can learn from the past. It is rich with our mistakes and triumphs. There are many simple inventions and methods we might be wise to revisit. I have a lingering disquiet that it might be a smart idea to have a few vintage handtools around that don't require any form of energy beyond my own.

In the past, because materials were more scarce and expensive, the things you had needed to last. Today's economy is based on short lived products. Some items are so fleeting in their composition they feel like a gesture of the products they are held up to be.

obsolescence . don't buy it

We may experience a bump here. Manufacturing culture relies on built-in obsolescence. But this is not environmentally responsible or fair to the consumer. Sadly many of the craftspeople who made a living by repairing things like shoes or vacuum cleaners are gone.

We have accepted as normal, "toss it and buy another."

At our landfill when we put things in fair shape next to the dumpster, unless my buddy Jr. is on duty, his sanitation boss wants it thrown in. "It's policy," the man says, like that explains anything. Jr. shakes his head slowly, turning to look at the mountain of landfill with a pained expression. "What can you do?" he shrugs.

Why not have a section at the dump next to recycling where folks can put used things others might wish to have?
That is how it was back in the day when everyone loved going to the dump!
The phrase "one person's trash is another one's treasure" surely came from this.
You can also put anything in good shape online through many "free stuff" sites.

One way to stop companies from selling us cheaply made junk is to leave it on the shelf.

Reward companies that make things well by purchasing from them.Ideally, companies should be responsible for what they make from beginning to end. Companies like Patagonia actually accept product back when customers are done with it. They refurbish some clothing to sell as used with a lower price tag.

The trend since the 1990s has been to produce lots of product cheaply to sell cheaply. Price was the bait to hook the consumer. Quality was compromised and then forgotten.

My parents grew up post-depression, when the country had grown extremely frugal to survive. Every hard-earned purchase was a big decision. Their generation trended toward buying less, keeping it in good condition and using it until it truly wore out.

Many baby boomers and generations beyond buy based on want, not need, chucking it when they don't want it anymore, useable or not. Big and more have been equated with wealth and success. We are on the tail end of this wild buying frenzy with resale thrift stores now a multi-billion dollar a year industry.

Life can only be understood backwards; but it must be lived forwards.

.... Søren Kierkegaard

Happily, a new trend toward small and accumulating less stuff is being embraced especially by younger people. Many grew up in over-stuffed households. Social campaigns have also increased awareness of child- or slave-labor practices in dirty, dangerous factories, producing unregulated goods.

If we agree to pay more for better quality things, then the makers can earn a living wage, and our purchases should be both safer and more durable. Living with less clutter, our carefully chosen stuff will be able to fit inside our new "tiny" houses.

Landfills can become our best resources as sought-after materials grow depleted. It would be prudent to sort and store like materials now to recycle instead of burying plastic bags of mixed garbage. Many common items currently tossed, like aluminum cans, should never go to a landfill.

what to do with other junk

Look online for your local recycling resources
(ehso.com/find_a_recycling_center.php)

Electronics . like old televisions, computers, printers, monitors, cell phones, tablets, iPads, toner cartridges, and batteries may be accepted at instore collection events (Best Buy, Staples, Walmart).

Hazardous waste materials . should never go into your regular trash. Pull aside for pickup during community collection events (old paint, pesticides, chemicals, cleaning fluids and other toxic stuff). Visit Waste Management **wm.com** to learn more.

depose of correctly please

Automotive products: antifreeze, fluids, motor oil, oil filters, gasoline, polish, and waxes
Batteries: home and vehicle batteries
Light Bulbs: Fluorescent light bulbs and compact fluorescent lamps (CFL)
Mercury containing devices: thermometers and thermostats
Paint products: oil-based, latex, and spray paints, caulk, wood preservative, wood stain, aerosol cans
Drugs or vitiams: pharmaceuticals of any type
Garden chemicals: pesticides, herbicides, fertilizer, insecticides
Sharps: syringes, needles, and lancets
Swimming pool chemicals: pH stabilizers, chlorine, algicide, etc.
Household chemicals: toilet bowl cleaner, shower/tile cleaner, oven cleaner, carpet cleaner, rust remover, etc.

Note that these are typical items routinely found in our homes that are too hazardous to throw in the trash. Something to think about.

Carpets . CARE **(carpetrecovery.org)** was created to promote recycling of post-consumer synthetic carpet. Utilize their online collection center map to find a location near you to turn old carpet into building or paving materials, insulation, new carpet, and all sorts of useful things. Carpets and rugs made out of organic materials may be used under mulch in garden areas or walkways. They work as top notch weed busters.

Packing materials . bubblewrap, packing peanuts, cardboard boxes. Save and reuse to ship stuff or donate it to a local business that ships.

Clothing . first offer to friends or neighbors. It's fun to get new "old" things. Donate to churches or Goodwill. Even tattered clothes are bundled and sold to recyclers to make new product.

Takeout containers . Plastic containers must be thoroughly rinsed out. Oily residue or food/grease left on the containers make them unrecyclable and may contaminate other items. Reuse them for lunches—or don't take them in the first place (bring your own container or ask for wax paper).

Hard plastics 3-7 . cannot be recycled due to lack of a market. Check the # on plastic containers to make sure they can be recycled (#1 & 2 are ok). Yogurt cups, butter tubs, oil bottles are not recyclable.
So if you buy it, figure out how to re-employ it.

MIXED . No Can Do. Try not to buy products with mixed materials, such as Pringle's potato chips (cardboard, foil, metal & plastic). Remove metal tops & rings on glass bottles if possible.

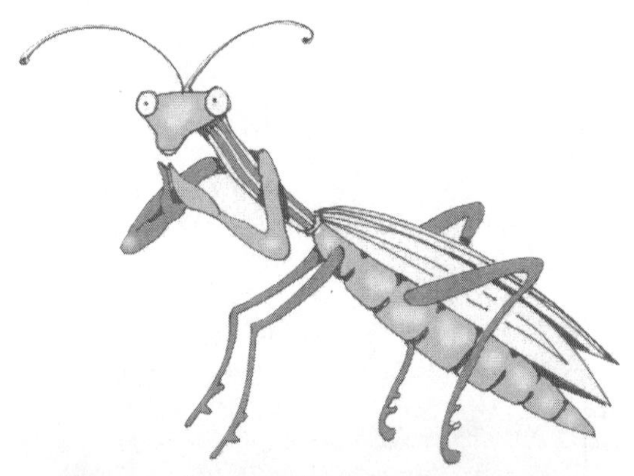

Please do your best to fix the mess.

always recycle :)

1. Aluminum
2. PET Plastic bottles (#1)
3. Newspaper
4. Corrugated cardboard
5. Steel cans
6. HDPE plastic bottles (#2)
7. Glass containers
8. Magazines
9. Blended paper
10. Computers
11. Lids
leave lids on plastic containers and Tetra-Pak boxes
12. Metal bottle caps
place caps in an aluminum soup can until it is half full. Crimp the top of the can so that the bottle caps are trapped.

can't recycle :(

1. Shredded paper
reuse as packing material

2. Colored or glossy paper
reuse as wrapping paper or packing material. Take magazines to shelters, libraries, or schools

3. Pizza boxes
reuse under garden mulch

4. Home glass
light bulbs, mirrors, panes. for proper disposal see: WM.com

5. Milk or juice cartons
because of the wax layer lining; reuse for crafts

7. Paper coffee cups
because of the plastic coating; reuse to start seedlings

8. Used diapers
whew. BIG stinking problem.

9. Plastic netting
birds & water animals get caught in it. Collect several into a ball for pan scrubbers, or make net bags.

10. Black plastic
black microwave food trays, takeout containers. DO NOT reuse—toxins leach out after one heating! **Avoid purchase.**

NOTE: Don't mix "Compostable" bioplastic cups, plates, and cutlery in with other plastics. They need sunlight and about 6 months to degrade.

mining the past

good old ideas for today

1. Keep a washable **cloth hanky** on hand instead of going through tissues and paper towel.

2. **Handtools** last longer and are easier to store than electric gadgets (hand can opener, coffee grinder, crank spiralizer, beaters, veggie graters, etc). Use no energy except yours.

3. **Preserve your own food.** During harvest season buy large amounts of produce to can, dry, or freeze. There are great online recipes for making anything from bread to yogurt to ginger beer. Preservative-free homemade food always tastes better. Plus you control the salt /sugar content, and may claim the proud DIY factor.

4. Keep a **rag bin.** Cut up old towels, t-shirts etc. Use for dust rags or to clean up. Wash them and reuse.

5. **Regift, trade, lend, give.** Back when things were expensive or scarce, little was tossed. Neighbors borrowed and returned things they needed. If someone loves something you are done with pass it along. We used to enjoy the best gifts ever from our friends Michael and Kathy, flea market aficionados, who had a house full of amazing whatnot. We would exclaim, "WOW, where on earth did you find this?!" Out there in the world of stuff!

Some amazing gifts we treasure from our thrift saavy friends are:

a bakelite radio that only got one station (greek music)
a vintage black velvet opera coat
with a white (fake) fur collar
a deco milkshake machine from a diner
and a white tyvek suit

We have adopted this tradition for gift giving. It is fun to search for a curious found thing and wonderful to receive it.

Always nothing you might expect.

use whatcha got

Our favorite game. Big reward! Inexpensive and gets rid of stuff. Before you purchase something you think you need, look around. So much of what we have can be adapted or repurposed saving time and money along with scoring you **green points.**

.... It can be as simple as pulling the empty cereal box bag out (as you place the box into cardboard recycling) to contain your lunch sandwich or other food. These are easy to rinse out and hold up well for several more uses and then into recycling please. (Cereal box liners are usually made of High Density Polyethylene (HDPE), #2 plastic.)

.... What is lying around, gathering dust... unused stuff in your way? It might be just the thing you need for that little project that has been on your mind.

.... Maybe you always wanted a little greenhouse. Are there some old storm windows with screens around? (Open the glass to screen in summer, so the space isn't too hot.) Look online for easy plans and give it a shot. (If you need old windows try the Habitat ReStore.)

.... Cruise Pinterest or the many maker sites online. There are loads of designs and ideas to inspire you. Things like old pallets are sought after as the basis for hundreds of easy projects, from inexpensive furniture to moveable raised garden beds.

This game will become part of your retooling.
Instead of thinking to just throw stuff out,
you will begin to see its possibilities.
Using materials wisely, waste not, want not;
offers you the basis for invention.
Instead of waiting for what you need to become available, puzzling out a solution with found stuff is fun and you get bragging rights.

necessity is the mother of invention.

.... Plato

(although sometimes I say "Invention is a mother.")

what worked . old is new again

Beyond culling materials from spent stuff or reusing things, our most valuable harvest from the past is the tried and true ideas that people have depended on for centuries. From stackable vintage glass refrigerator boxes instead of short-use mountains of odd shaped plasticware tops and bottoms that don't fit, a look to the past will offer endless smart solutions.

Biochar . is a system originally practiced by early South American civilizations that traps carbon producing "green coal." Biomass waste releases carbon as it decomposes. Things like wood or peanut shells are burned in a kiln by pyrolysis, an airless burning technique. This creates biochar, a soubriquet green coal, which is dug back into the ground in order to lock carbon into the soil. Removing agriculture and forestry residual biomass carbon by sequestering it underground also anchors soil nutrients by adding lost carbon drained by industrial agriculture. Biochar could lock away excess climate warming carbon for at least 100 years.

moving forward by looking backward

Early people sheltered in caves or created burrows to protect and insulate them from the elements. A return to soil as an insulator is most visible as the base for green roofs hosting gardens and carbon collecting plantings. Berm or sod houses partially build a portion into a hill or literally use dirt pack as a wall material.

Green boxes are a great alternative to toxic formaldehyde laden commercial coffins. A biodegradable pod or a simple pine coffin holds the deceased. A tree may be planted above. As the body degrades it provides nutrients for the tree to grow. Stone graveyard markers could be replaced by beautiful forests.

Traditional homes were built to face south for heat in winter utilizing the sun (passive solar), one room wide with windows placed for top air circulation, high ceilings, roof overhangs and deep porches to keep cool in summer. Energy efficient LEED construction returns to these ideas.

Enough is enough

In evaluating our ballooning excesses during the last half century, fueled by an attitude of self entitlement coupled with greed, the question is beyond what is enough for each of us? We should ask:
What is enough when we consider the needs of all living things and the health of our earth?

For those of us in developed countries accustomed to a high level of comfort, some readjustment to less is in order. For those struggling for day-to-day survival in third world countries a step up in comfort can also benefit the health of the planet by removing the need to clearcut forests or kill endangered wildlife.

Climate scientists, world-wide, agree the "greenhouse effect" is a result of human consumption and expansion over centuries.

greenhouse effect:
when our atmosphere traps heat radiating from Earth toward space reflecting it back causing an increased warming cycle of the planet.

More people, industries, products, food production, air and car travel, cities, etc. result in increasing issues of waste, carbon, pollution, and land use worldwide. We are both the problem and the possibility for real change. Let us take it home, you and I. What is global starts local. We can start the easy changes contained in these pages that will save each of us, money, aggravation, and time to appreciate what is truly precious in this world.

"The world is changed by your example, not by your opinion."
. . . . Paolo Coelho

Here's a game. Let's call it **Wretched Excess.**

Take a week to write down all the things you note to be obscenely wasteful practices...

...like stores that have their door wide open pouring AC into the street during the heat of summer to be welcoming. Next, consider if there is a politic way to prod a change to greener thinking. The first time I encountered the above example, our family was visiting NYC. It was 90 plus degrees, and our teenage daughter was sucked through the gaping glass entrance into an H&M. Cathedral ceilings rose through the center of the massive clothing store to four floors up. As our kids got lost among endless racks of merchandise, my husband envisioned hundreds of dollars flying out the door while I envisioned the end of the world.

Several years later in Savannah, Georgia, our family passed a small eccentric vintage store, once again with the front door wide open, blasting their AC out onto the smoldering street. This time after a friendly comment on what a neat store it was, I asked the proprietor, "Doesn't it cost you a lot to leave your door open like that?" He shrugged, saying, "Yeah, but it brought you in, didn't it?"

There is a bunch of stuff in play here; businesses, organizations, communities, and individuals feel entitled to act for short-term gain. People want to continue to do things, simply because that is how they have been doing them. I recall a neighbor, an older waterman, who had worked the Chesapeake Bay for 50 years responding to new limits set on catching blue crabs.
He puffed out his chest and declared,
"My daddy plundered the bay,
his daddy plundered the bay,
and by God, Ah'm gonna
plunder the bay."
This type of thoughtless logic is why there are few blue crabs left in the Chesapeake Bay and we find ourselves facing the biggest global crisis humankind has ever faced.

Back to our little game of cathartic power, this one addressing junk mail, solicitations, glossy marketing postcards, and unwanted catalogs. As an environmentalist, I get piles of fat envelopes containing beseechments for contributions from organizations concerned about the earth. There is a sad irony in that, usually causing me to curse aloud, unnerving the cat. Think of the printing, postage, and any production costs for this harassment that might have gone to supporting their cause. As you toss it into recycling remember to separate any plastic windows from paper envelopes.

your move:

Call the organization, tell them you want to opt out of all their mailings and explain why. They may listen.

You can also decline online.

To curtail all the random junk mail coming your way, consumers can register at the Direct Marketing Association's (DMA) website: (DMAchoice.org)
Opt out for a processing fee of $2 for a period of ten years. Registering online is the fastest way to see results. DMAchoice offers consumers a simple, step-by-step process that enables them to decide what mail they do and do not want.

game on!

Call 1-888-382-1222 from the phone number you want to register for the Do Not Call registry. YAY!

Unsubscribe!

Take some time and release yourself from spam email by finding the unsubscribe link. They are stealing your mental energy and real energy with their practice of consumer harassment.
Computers and the web guzzle electric while you hit delete over and over and over.

say enough already
to junk mail, phone calls & spam

billify & vilify

Most utility providers, cable companies, and financial institutions offer the option to receive online statements instead of mailed printed statements. You just need to log into the company's website to change your preferences. It may still be beneficial for your records to create a ledger in Excel or just a Word document detailing each payment amount, date paid, and to whom. A single sheet could be printed out for a monthly hard copy file.

I get more spam than anyone I know.
. . . . Bill Gates

Opt out:

You can put a stop on unsolicited commercial mail, credit card, and insurance offers. If you decide that you don't want to receive prescreened offers of credit and insurance, you have two choices: You can opt out for five years by calling 1-888-567-8688.
To opt out permanently online go to (optoutprescreen.com)

more fun :)

Stop the relentless onslaught of (glossy toxic ink) catalogs! Call the catalog company or write return to sender on it and send it back. That usually gets their attention.

Now you can look forward
to just getting nice letters from your aunties
and your voter registration card
in the mail.

Wretched Excess 2.0

Packaging . Problem, you like or need the product, but it comes smothered in unnecessary wrapping. Let the company know. Your concerns as a consumer are the basis for their success.

Medical & Dental waste . Each patient visit results in a pile of single-use plastic, tossed into the medical waste container's plastic bag. Offer your concerns to your doctor or dentist. You are both his customer and patient. Ask if there are alternative products to use. People are seduced by the next "better" product offered to them.
I am curious if we are safer using all the plastic medical paraphernalia than when metal tools were sterilized and surfaces and hands were washed between patients.

U.S. hospitals create 5.9 million tons of waste yearly, 30 percent of it generated in the O.R.(operating room). New green waste treatments using autoclaves, microwave units, hybrid steam treatment systems, and other steam-based technologies, combined with mindful attention to reuse some materials has resulted in an 18 percent reduction in landfill medical waste.

Biodegradable natural rubber latex or nitrile safety gloves are available for use instead of blue plastic surgical gloves.

Small actions count . You can switch to: plastic-free dental floss (biodegradable/compostable). Over 3 million miles of dental floss are purchased in North America every year. That's the equivalent of reaching to the moon and back over 6 times.
Bamboo tooth brushes are a good step forward, although most of their bristles are still petroleum based or pig hair.

Share your concerns
with your doctor and dentist.
This is a huge waste area.
They can ask for better.

HORRORS!
Don't even think of buying/using it . ever

1. Cling wrap
Invented in 1953 back when plastic was the go-to for everything. It is non-recyclable, non-reusable, and loaded with PVCs making it potentially hazardous due to the fumes released when its structure is tampered with through things like heating. Use waxpaper, aluminum foil, a plate, or a reusable container with a top.

2. Flimsy plastic bags given out willy-nilly
Yes, I am talking about the bags you just brought in with your groceries in them. Sure, you may reuse them in your tiny wastebaskets, but unfortunately these bags wreak havoc on the machinery at recycling plants by causing jams, blow out of the landfill to lodge in trees, look like yummy jellyfish to turtles and ....well, please just stop taking them. Use your cool market bags; so other shoppers may follow your lead.

3. Paper plates, cups & plastic utensils
After these things are defiled by barbecue and wedding cake, they can only be thrown away. Why not rent real dishes for the party or wash the dishes you use. Faux plastic silverware costs more than miscellaneous homeless silverware at the Goodwill or even bulk aluminum cutlery at Sam's Club in the restaurant section.

4. Anything made of Styrofoam
As common and disposable as Styrofoam is, there is no way to recycle it. It's also highly flammable, making it a hazard for recycling centers. Happily some states are or have banned single use polystyrene.

Just walk past it
like you don't even see it.

drives you crazy

In the past, people lived near work so they didn't have to spend much time or money to get there. Car sizes balloon up or down depending on gas prices. As we reach the decade of 2020, the market offers very tiny cars, environmentally responsible electric cars, or a parade of BIG pickup trucks, SUVs, and luxury sedans. This reflects our schizophrenia relating to consumption and climate change. Most American households have 2 or more cars. The majority of cars on the road contain one driver. According to the U.S. Census Bureau, the average one-way commute time is 26.1 minutes. If you commute to a full-time, 5-day-a-week job, round trip; that adds up to 4.35 hours a week and over 200 hours (nearly nine days) per year.

an average U.S. commute is 4.35 hrs per week

Before carpool was a word, people rode to work together because not as many people had cars. Gas was considered an expense, so people combined errands and took fewer trips. Even a return to less driving is a big step forward. Transportation recently replaced power plants as the number-one source of U.S. carbon dioxide emissions. Trucking and air travel (diesel and jet fuel) are a large percentage of the rise, as more people fly and order stuff online.

Shocker . Start a car log, with destinations, time in car, mileage & gas dollars spent. We drive without considering the real cost to us in time or carbon emissions expended. It's pretty wild to look at that log after even a week. Give yourself brownie points for passengers and combined errands!

You can also see how your car is doing on mileage. Make sure your tires are inflated to the proper level to get more miles per gallon.

A sample log to use is in Action Journal section page 164

Game on . Stick shifts in cars used to allow drivers to conserve gas by going into neutral cruising downhill or up to stop lights. There are ways to eco-drive saving money/gas/emissions. Hypermilers push fuel efficiency to the limit. Correct tire pressure on your car saves 10% on your petrol. Reduce your top speed from 75mph to 65mph. It can cost you up to 25% more in fuel to drive at 70mph compared to 50mph. Drive smoothly, not crowding the car in front of you. Turn the car off when paused; idling your engine increases emissions by 13%. Windows up, roof racks off, and keeping your car load light saves gas. However, windows down to use less AC will save more gas during summer. Hypermiling can also improve your alertness, adding another level of interest to your trip.

air traffic control

When gas prices go above $3.00 people think before they drive, planning more efficiently about how that gas will be spent. Vacations may become staycations or at least going to places closer to home. Getting on a plane has become commonplace, even with all the hassles of flying in a hypervigilant world. In fact air travel is surging just when we need to seriously curtail it to meet reduction goals needed to offset catastrophic climate change.

Every round-trip transatlantic flight emits enough carbon dioxide to melt 30 square feet of Arctic sea ice

Flying should be considered only when neccessary, or at least as a rare special treat. Jetting to Aruba for the week does not fall into that category. If you must fly please note:

A shorter flight's carbon intensity can be higher than longer ones. Planes take a huge amount of energy just to get off the ground. All air travel spews particles, nitrogen oxides, and sulfates trapping heat that is especially damaging at cruising altitude. Sadly, new technologies are nowhere near any solutions for greener aviation, so curbing emissions is our responsibility in choosing how much we fly. Rationing has been proposed, or an individual cap on air travel that people can trade with each other as a tool to help reduce the impact of flying.

Global tourism now accounts for 8% of carbon emissions.

**We need to rethink BIG.
Enough might be a better comfort level
to aspire to in many areas of our lives.**

In the past most homes were a reasonable size, easy to heat and care for. People selected where they lived based on cost concerns and proximity to work or family. With the advent of cars came suburbs and sprawl. Before the real estate bubble and crash in the mid-2000s, greedy developers were slapping together poorly built massive McMansion developments with abandon. Yards swelled to wide pesticide and fertilizer infused expanses. More fossil fuels are required to mow a big lawn, go to work from and run these large, inefficient homes.

Charming historic downtowns became ghost towns as retail moved off highways to shopping malls anchored by department and chain stores. Many of these are now shuttering as consumers concentrate their business at big box stores or buy online. At present, we are left with struggling downtowns and a few big retail players to buy from. The trend to buy BIG quantities to save a buck is ubiquitous.
But purchasing more requires extra storage concerns like increased energy usage in refrigeration or even food waste as shelf life expires. Conversely there is more, and often redundant packaging to dispose of involved in purchasing large packaged goods. Are these huge stores offering more than we need really a good thing?

There is a trend to declutter as people are overwhelmed by their stuff. Some elderly neighbors of ours needed to move to a small simple space in a facility. Their health was tenuous, making it dangerous for them to drive to doctors or even to get food. While friends pitched in to help them, we couldn't all the time. They became upset at the thought of leaving their overstuffed house." What will happen to all our stuff?!" they cried, literally imprisoned by their possessions. Studies have shown that extreme clutter or symptomatic hoarding creates anxiety. Less is better and a goal to work toward. We are in editing mode at our house. Feels refreshing.

Too much is just too much.

A tomb now suffices him for whom the world was not enough.

.... Alexander the Great's tombstone epitaph

A few notes to encourage you...

We are lucky to live in a world of abundance. Maybe this entitles us to the superpower of giving. You will find a comfortable "enough already" when you look around and feel gratitude rather than stress with what you see.

The big phrase in our family, courtesy of my mom, Jackie Fritch, is

"Moderation in all things."

She was a very smart woman and this does hold true in many areas. The first step is awareness, the second step is setting doable goals. Better to move forward moderately and mindfully than to feel overwhelmed or paralyzed. Each of us has different variables influencing what we can do to lighten our personal impact on the environment. Work within your constraints and be proud of even your tiniest successes.

Case in point: Our home is a BIG old farm house (circa 1790). We replaced windows, tightened doors, insulated and continue to do what we can to make it as efficient as possible. This has taken years and lots of elbow grease. We have worked creatively with what we had, or could get, to make improvements. In winter we hunker down to a few rooms (we get small), in summer we depend on a good breeze from the bay and run the AC only on super hot nights. Luckily the Mid-Atlantic offers 9 months of fairly moderate temperatures.

We commute about 35 miles to work, riding together if possible in our 2008 Subaru wagon. (Sometimes driving through a high tide and always down a bumpy dirt road- so a low base electric car is not an option). Living increasingly green is an ongoing process. Count your successes as you move forward to retool. Doing something is good.

imagine positive

. . . . . developing a self nurturing mindset

tools

play & invention

anticipation response

a perfect world

living a joyful life

This book is the result of Paul Hawken

After my husband and I attended an uplifting lecture by him at Loyola University that we drove 3 hours to see, there was a question and answer segment. First let me say that this fellow is sharp, witty, and an absolute treasure. Paul can make 150 people feel like they are sitting with him in his living room as he hands out the complexities of climate change in easily digestible bite sized triscuits. The lecture was on his wonderful book and project; *Drawdown*. Updates and more information on both of these are available through (drawdown.org), including positive strides by scientists, scholars, and organizations to combat climate change worldwide, free of political constraints. I was inspired to be useful too. "How might I help, as an artist, someone who is not a scientist," I asked, as the last questioner at evening's end. He smiled pleasantly and replied, "Well, I don't know you. But I bet you will think of something."

I began to look at my skill set and how I might employ it as a creative person to help work for a solution. I played with ideas, questioned the status quo and invented less impactful approaches in my daily life. At some point, it dawned on me. The fastest and best way to make change is through each of us. I retooled our home and business, and began making my own toothpaste, mouthwash and cleaning products, becoming more intrigued the more I learned. Why were we so at odds over climate change when it just requires us to take a hard look at our actions and work together to reverse their effects. How could I get beyond our current counter-productive culture of division and mistrust to start this critcal conversation with as many people as possible?

play & invention

The key is to imagine what you can do with your
skill set to accomplish your goals.
Dare to be inventive. Explore, learn and play.
By the way, making your own toothpaste is
absolutely fun and better for your teeth.

As a writer and illustrator a book was already forming in me. Ideas and information needed to be conveyed so that other individuals would motivate to join this time-sensitive movement. To get beyond politics and conjecture, to tackle the many angles of reaching a greener path, a handbook came to mind.

Most important in retooling your approach to using our planet's limited resources is changing your perspective to a long term view. In choosing to live your life responsibly for the health of future generations of earth's creatures, your small actions carry more weight and self-gratification. When you borrow something from another it's important to return it in the same or better condition than it was lent. To walk lightly during your lifetime, leaving Earth as you found her.

This is a notion subscribed to by "ancient" peoples that lived close to the land. They understood that in compromising or overusing earth's resources they would ultimately threaten their own survival. The land and nature that sustains modern cultures is far from our manufactured mindset. Returning to a more grateful outlook, realizing the real cost of our conveniences, helps one to live more sustainably.

I experience a small moment of awe
every morning when I turn on the shower
and feel the luxury of hot water
pouring out like magic.

play with your food

Breakfast can be boring.
Some people have resorted to
unwrapping sugar-infused "breakfast bars"
full of implied nutrition to choke down with copious
amounts of coffee. Horrors!

Find your inner little kid.
Have fun when you eat. Make more so that every
meal has extra to grab when time is short.
Here are a few ideas, but once you get your
spatula primed, the sky is the limit.

. . . . flap jackies

Make a pile. Sure you can have them with mad dog gravy (maple syrup) but that is just old school.

mo better: Use them to roll up stuff like a burrito. Wrap in wax paper and you are out the door.

fillings: nut butters & banana
fruit, plain yogurt, walnuts, honey
or scrambled eggs, cheese, mushrooms

. . . . oatmo better

Great hot or cold. Bring a cup of water or so to a boil. Throw in oatmeal, dried craisins, figs, walnuts, a chopped banana or apple and unsweetened shredded coconut.The boil makes it all creamy.

. . . . or cold brew

Put oats etc. in a jar with almond milk or (any milk) to brew in fridge overnight. Perfect in the morning. Cold brewed coffee or suntea are also treats!

Oatmeal and pancake mixes usually come in simple uncoated cardboard boxes, that can go right into recycling.
Or create your own mix. . .

mix it up

Gather dry ingredients together as you like them in a covered container to save time in preparation. Combine cereals in a single container to make them more interesting (Grape Nuts, Bran, shredded coconut, walnuts or almonds, craisins, hemp hearts, rolled oats). I use a covered plastic pitcher; giving us more room in the cupboard and a quick pour.

GIFT IDEA! Make a fun label to dress up your mixes to give as a useful present. This also passes on the idea of less packaging, less preservatives, less boring and way more yummy.

make it up as you go

You get a gold star and bragging rights
every time you reuse/repurpose something that
would have been tossed.
In this game,
that is a **save!**

Next are a few **"Makers Ideas"** to enjoy making on a rainy day. Handmade gifts always mean more than factory produced stuff.

Plus there is a bonus.

There is the sweet joy
you get in the doing of it.

Creativity is intelligence having fun.

.... Albert Einstein

To practice any art,
no matter how well or badly,
is a way to make your soul grow.
So do it.

.... Kurt Vonnegut

Art is good for you. It allows you to slow down. Give yourself the license to experiment. Remember that you might have thrown this trash out; instead, you are making something. If you think you don't have time to play, consider that this is also a valuable exercise in freedom. Freedom to take time, to let your mind wander and freedom to enjoy yourself in the doing.

bottle caps . True, it is frightening to see the pile we have in the bucket under the bottle opener in our house. I blame my friends (somewhat). You can recycle them secured inside a metal can, however we have found them to be quite useful for various projects. An artist friend flattens each one out with a hammer and then places them color side up in rows (like scales) over wood shapes for fantastic works of art. Cover table tops or mirror frames for durable decor.

An easy quick project is to make bottle cap magnets, pins, or earrings. Glue a pin or magnet on the back and glue an image in the bottlecap (anything, from a small picture of a pet or friend to artwork). Or break out the glitter and fill the cap with a glue/glitter mix. Place a cut out inspirational word over that like "peace."

Make earrings in the same way, but first punch a hole into the side of the bottle cap, thread wire loops through and cover with the image you choose. Use a small nail and hammer to create the holes in the bottle cap (Put some scrap wood underneath to protect your work surface). Purchase earring hooks from a craft store and attache with a jump ring.

. upcycle it .

threads . Think of anything that is cloth as fabric to be used. Upcycle an old linen table cloth into reusable, washable napkins, dish towels, or even make clothing. Old towels may be cut up for dust rags, or backed with a sturdy fabric and stuffed with plastic netting like onions or lemons come in for a useful dish scrubber. Give it a wash when dirty, make another when it wears out. Commercial sponges harbor germs, last only a few weeks and are typically made from polyurethane, a petroleum-based material that can't be recycled or composted. Switch to using your handmade beauties. Give them as a gift with organic dish soap.

old t-shirts . Make stretchy carrier bags from old t-shirts. Remove the neck and sleeves making the handles and bag opening. Sew the bottom shut or cut ribbons 5 inches up from the bottom for a no-sew. Pull one strip from front and another from the back, knotting to close the bottom of the bag.

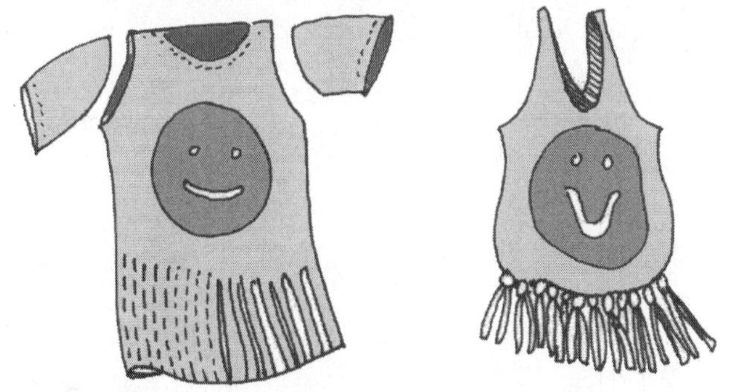

stockings & tights . Make great dusters, if you slide a rag inside, the webbing grabs dust. To use old stockings to filter liquids, oil, or lumpy paint; stretch a section over a container and knot the open end.

milk & juice cartons . Despite the fact that cartons are made of 85 percent paper, 15 percent polymer, and are largely recyclable, not all municipal curbside programs accept them because they fall under the material category of paperboard lined with wax. Purchase these liquids in recyclable #1 PET plastic (polyethylene terephthalate) or #2, HDPE (high-density polyethylene) instead—and just make creations with cartons:

birdhouses . Staple the top of the carton shut, paint it a jolly color, let it dry, and cut a front door hole in its side. Next, cut several small holes in the bottom of the carton for water drainage, a small hole through the top of the carton to thread string or wire through, and hang the finished birdhouse in a tree. **Tweet!**

votives . Cut the gable top off the carton. Stencil or freehand designs to cut out on each side. Place a candle inside to line your walk or doll up a summer party.

soup blocks . Great for freezing soups or smoothie fruit. You don't have to thaw out the whole container if you only want a single serving, just saw off what you need.

in the garden . remove tops for reuseable seedling starters or to make plant collars: remove the tops and bottoms from cartons. When the ground is soft, push the collars into the soil, around the plants to discourage grubs and cut-worms from attacking young tomato and pepper plants. The collar also retains water to stay just around your little plant instead of draining away. Pack mulch into the carton around the plant to retain moisture and stop weeds.

melt it . Last bits of candle wax or bits of soap can be gathered in an old coffee can and placed in a pot of boiling water to make a double boiler.
Gather small milk cartons for molds with the gable top opened or removed. For candles buy some wick at the craft store, lay a stick across the top to attach the wick to, pull straight down and tack the other wick end into the mold bottom. Once the wax is liquid, pour it into the molds for new candles.
With soap, once melted, pour into the base of the carton for a new block of soap. When the soap or wax has cooled and hardened peel off the carton paper. Voila! New soap and candles for a bath!

organize it . Pull 6 cans aside, wash, and remove paper labels. Dull any sharp burrs around the cut edge of the can using a sanding stone or file.
Line up three cans, measure, and cut a 1x6 board to length (8"-9"). Paint cans and wood an upbeat color or leave natural. Using a brad-sized nail and hammer, or similar-sized drill bit, put a hole in the top portion of each can. Use a strong glue like E5000 to put a bead down the side under the screw hole, push tight to the board and screw in place with a short, small screw & cap. Install the remaining cans in place on both sides of the board. Attach a handle to the board with wide topped roofing nails. (Upcycle the strap off an old handbag, use a cupboard handle, or cut a piece off a tired leather belt.) This makes a great tool, picnic silverware holder, or garden organizer.

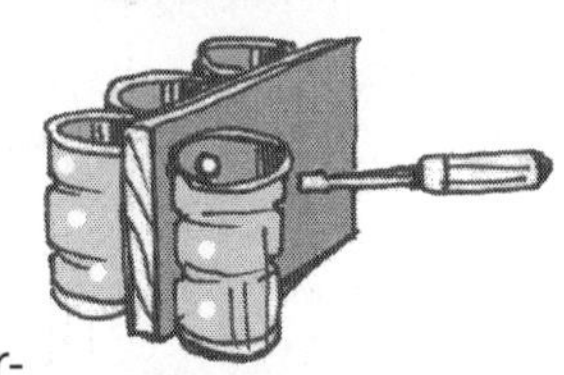

or . Use screw-top glass jars to sort hardware or other small items. You can punch a hole in each lid and attach the metal lids to the underside of a shelf in your workshop or closet. Screw each jar back into its top. This gives you more room on the shelf and makes it easy to see each jar's contents.

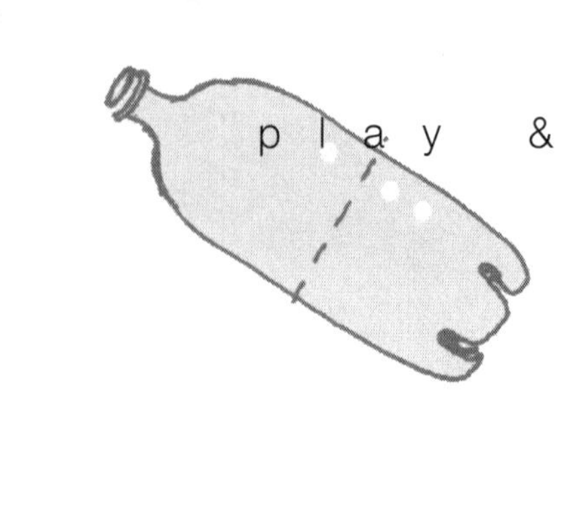

wine & beer bottles .

Create a vase or drinking glasses. Use a glass/tile cutting tool/kit to cut neck/ shoulder area off. Place some 60-grit sandpaper in a shallow bowl and cover with about an inch of water.
Sand the edges using a combination of circular motions to dull any sharp edges of your new glasses. Ensure that the portion you're sanding is submerged in the water. Sand the inner and outer lips well since these will both make contact with your mouth. Finish the sanding off with some 120-grit sandpaper. You can use the top halves of the upcycled bottle as candleholders for tapers.

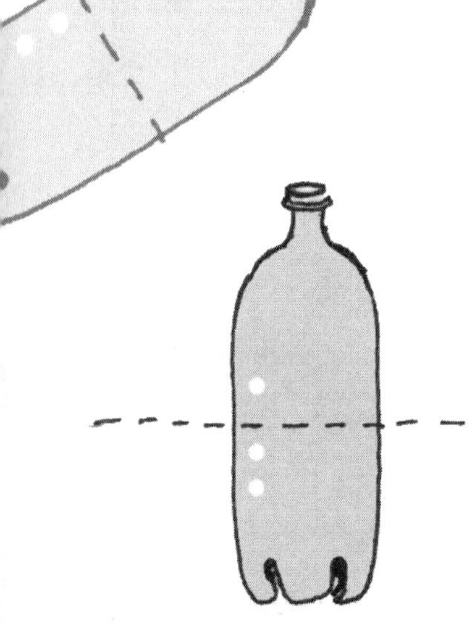

liter plastic bottles .

Make a self watering planter . Use a utility knife to make a slice mid bottle. Use scissors to separate the halves. With the cap on, invert the top portion into the base to fill with soil. Tuck in a seedling. Lift from base and water the new plant also adding water in the lower half to cover the bottle neck. Remove the cap and replace the top with seedling inverted in base. Set in a sunny spot.

Plastic bottles are a popular material for all kinds of amazing upcycle projects from constructing house walls to actual boats that float. They are also useful for brewing ginger beer. Explore Pinterest online.

aluminum cans .

Wearing safety glasses and protective gloves, cut open and flatten an aluminum can using tin snips or a Dremel MultiTool and cutting disk. Press into the reverse unprinted metal side with a pen or pencil to create raised designs. You may add color using sharpies or a wash of black acrylic paint—accent by removing the color (after it dries) from the raised areas with a scouring pad. This thin aluminum is similar to the copper foil you can purchase in craft stores. Turn your creations into ornaments and jewelry, or frame for wall art. Of course aluminum is a top thing to recycle too, so save your scraps and do so.

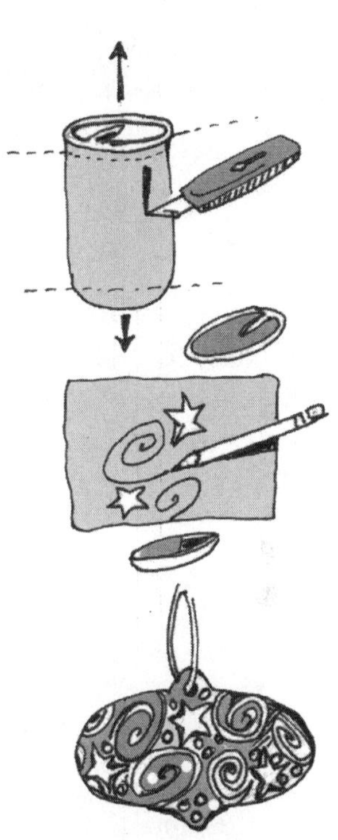

Think of anything that might normally get tossed into the trash or recycling as free material to mess around with.

Take over the kitchen table, involve your kids and family, or invite some friends over and prepare to be pleasantly absorbed in creating. It might even be fun to meet regularly as a **"Makers Group"** in the tradition of a book club. Each member could turn up with some sort of "trash" as a prompt. The challenge would be to puzzle out an upcycled creation. Envision great music playing in the background and a comforting beverage.

party! I have had craft activities as a side at parties where there are a variety of ages. This allows people to come together in an opportunity to talk and create.

Anticipation response

Note proactive ways you can easily operate
to lighten your own environmental impact and
inspire those around you to follow your lead.

When the waiter offered me a polystyrene box for my leftover dinner and I asked for two sheets of wax paper instead, something happened. As I folded my extra fish & chips in, our bar mates on either side watched intently, then aha expressions lit up their faces. They had realized, "We don't have to accept the status quo!" They each commented approvingly. Clearly, it's time to think outside the polystyrene box.

Big companies make their money by offering us "convenience."
It is their marketing plan
to hook us in a cycle of use & lose.
So just refuse.

Our Sierra Club printed nice little notes to leave behind in area cafes and restaurants encouraging them to switch to biodegradable containers. There is tremendous momentum to do so spurred from the grassroots. Businesses have caught on that being green is not only important, it is a powerful marketing tool. But while it's great everyone is getting on the bandwagon, one must wonder are they lip-syncing or really singing a new tune? "Biodegradable plastics" still take 190 days to begin to break down and can compromise the recycling stream when mixed in with fossil-based plastics. I called a company making plastic bags from potatoes to try and get some clarity. They did not offer the crisp definitive answers I hoped to hear. In fact, after some fumbling around they said they could not speak further. During a trip to Baltimore for a wedding and breakfast at the hotel's buffet, I noticed that the great piles of plastic cups piling up were biodegradable. As a waiter wedged them into a massive black garbage bag I showed him and explained, "Biodegradable plastic needs to go into it's own waste stream. It requires sunlight to break down." He just looked at me wearily and stuffed more in the bag.

The best proactive step that will guarantee real change in the mountains of takeout containers we generate;
simply bring your own container.

If I cannot do great things,
I can do small things
in a great way.

. . . . Martin Luther King Jr.

Pre-think your day.
Pack what you need to make your life convenient and environmentally responsible.

My husband, John, pulled over to help a couple that was having car trouble. Their vehicle had stopped working, leaving most of it protruding out, blocking the road. John jumped out to wave other drivers around. "Let me help you push your car off the road," he called to the couple. The man shrugged. "We can't do nothin' with it. This car is totally electric and it won't move unless it is on."

I took a ride with a nice friend and marveled at the beautiful, sleek design of his new car. It even had a push button to start it instead of a key. "But can't anyone start the car with the button?" I wondered. He held up his car key. "Nope. You have to have the key, but you can leave it in your pocket. The Bluetooth communicates with the car so it will turn on." Increased automation or complexity makes more stuff that can break.

Question convenience.

We stayed at an Airbnb with electric wastebaskets that opened when you pressed a button, a machine that played fan noises instead of a fan, and a coffee maker so complex we never did figure it out.

I am not a total Luddite, but all this dependence on technology and external energy sources makes me nervous. My truly happy place is driving my 1998 Rav4 soft top. (Actually it is a "no top," because once I wrestle those old zippers open, the top is off for the summer.)

The windows are crank, fueled by elbow grease.
The cassette player in it works great
(Just as it did, pre-CD, pre-iPOD, pre-streaming).

Companies need to get an environmental conscience.
That would be more convenient for all of us.

An escalator can never break: it can only become stairs.

.... Mitch Hedberg

Our economy is based on racing out to buy the newest, shiniest thing. That means working harder to afford it and tossing away the old things. It is a ladder into an empty sky of excess. If the thing works fine, maybe we don't need a new model or an update.

Who is leading us in this madness?
Is convenience ultimately all that great for us?
It might be better to just get up off the couch and do stuff.

A new convenience product can represent a tradeoff of a personal skill for technological dependence. We forget how to do things or worse never learn how, like, remembering phone numbers or writing a letter by hand without spell-check. Important human traits like patience, perseverance and problem-solving are waning. The simple notion of using your mind and body to try and tackle an issue is a rarity.

People are forgetting how to read maps. GPS is great when it works, but there are places where it goes haywire so a map is still not bad to have around.

We are increasingly dependent on devices that are not totally under our control or fully reliable. Many gadgets are made cheaply. The more complex they are, the harder and more expensive they are to fix. Yearly upgrades feed into the notion " just toss it and buy another."
Address books, calendars, calculators and spell check, movies, games, weather, treasured photos, banking information, and contacts are all in our smart phones, with slim back up should the phone meet a bitter end.

As gadgets get smarter,
what is happening to us?

A Joyful Life

There is an old adage that encourages:
"find joy in the work you do."
There is pride in doing a job well,
tackling a new task or accomplishing a goal.
The skill set you develop is your super power.
"Find joy in the work you do
and you will never be at work."

You, my friend, may note, as you retool to use green products and services, a deepening feeling of calm pleasure. I believe most people are aware that the daily excesses of our current lifestyle are not sustainable. As a culture we have become addicted to all the "conveniences" marketed to us. Most conveniences are more of a trade-off for something else. What do we do with the extra minute we get from a computer turning on the sound system in our automated homes?
Work overtime? Watch more reality TV?
I think I would rather fold laundry.

Finally, there may be a backlash to the tsunami of devices stealing away tasks that we have done well for centuries. The DIY (Do It Yourself) movement is the first counter-wave. This trend will be good for us. Our hands will once again be utilized as our greatest tools. The twin epidemics of depression and anxiety may quell as once again we become our own keepers.
It is good to feel useful.
It feels nice to do something.

**Happiness is not
something ready made.
It comes from your own actions.**
.... Dalai Lama

It is a simple thing to go green.

You buy things that are healthy for you and will last.

You depend on yourself to provide proactive conveniences.

The latest thing is not necessarily the best.

What works well is worth keeping.

Look to nature for inspiration and worthwhile invention.

Allow yourself to do joyful work.

Cleaning, cooking, driving, gardening, washing,
and the other tasks I do to maintain myself and home
offer a quiet therapy.
The yogis encourage us
to meditate on the beauty of simple tasks.

Sweep the floor, do it well,
enjoy the movement of it.
Empty your mind of clutter.

in a perfect world

Things go well, everyone is happy, content,
and the planet is at peace.
This is possible
only in a giving, empathetic, forward thinking,
responsible world.
We were given the gift of invention.
Altruistic invention and a new mindset
are in order.

We live on a small island in the Chesapeake Bay. It feels like a perfect world to us. Sure, there are challenges. If there is a storm it is full tilt, flattening the garden, throwing chairs off the porch, and blowing open doors if they are not locked. You cannot open the driver's side car door and passenger door simultaneously or any papers will fly across the marsh. There are waves of bugs. I have made peace with the black snakes that drape their skins from the rafters in my barn ceramic studio. We deal with a rising sea that carves away the shoreline, leaving tree roots bare and our well salty.

There is a narrow dirt road that winds out across the marsh to our driveway, crossing a culvert where the tides rush in and out. I keep a tide clock in my car and refer to it often especially during periods of the full moon and king tides. Driving through water on our way home sometimes is just part of what we do. It is sort of amusing to peer over the hood to see crabs swimming out of the way as the car whooshes forward with a little wake fanning out behind. We and many others are already living with the effects of climate change. Decisions we make are based on how to prepare best for what we may face out here in years to come.

This means it is an exciting time for invention. We will need to learn to live with rising sea levels and wild weather. We are figuring out how to batten down the hatches in creative ways. An old Chinese saying is "May you live in interesting times." We do.

Tropical storm Sandy took us by surprise, flooding much of our area as it roared up the eastern seaboard slamming New Jersey and New York. Our flooded road was impassable for a week, marooning us on our tiny island hill. It was forced quiet. No news or radio, a book to read only during daylight. I wandered around collecting a flock of duck decoys that had floated in and lost themselves in piles of flotsam. Some of our tin roof sailed off into the field, but generally everything was ok.

We walked down to the foot of the driveway to see if the road was passable. It was...with a boat.

I have always wanted a car that could drive through water. The only fan letter I have ever written was to Elon Musk imploring him to offer such a vehicle. There were lake cars in the 1950s that puttered around fairly successfully but there are few options today.

good ideas whose time is now

amphibious cars

1. When will someone design a car that drives through water? Think of the many people who could have saved themselves during Katrina or the Texas hurricanes. Most of the deaths resulting from hurricanes and cyclones are drownings.

Cities like Miami, Florida, and Norfolk, Virginia, already dealing with sunny day flooding during high tides could certainly use such cars. Venice, Italy, and many other places worldwide where the sea level routinely floods roads would benefit. Think of the big market just waiting—***call me!***

**Nature is the source of all true knowledge.
She has her own logic, her own laws,
she has no effect without cause
nor invention
without necessity.**

. . . . Leonardo da Vinci

Nature operates in small, decentralized interdependent systems. They work flawlessly intertwining all things in a precise dance of life. We have created the opposite, centralizing our power centers instead.

Every time there is a devastating weather event or fire that sweeps a community away, we have an opportunity to rethink how to better reconstruct it. Nearly always the electric grid supply is first thing to be severely impacted or compromised. After Super Storm Sandy, we sufficed with candlelight. I always keep blocks of ice in the freezer to make it run more efficiently, so we put some in the fridge to keep it cool. We have a wood stove for heat and a gas stove (propane) so losing electricity for 10 days was not as difficult for us as it was for the majority of people whose lifestyles count on constant electricity. Big electric companies want to control our energy consumption so they can control our dollar.

What if each house or community had its own decentralized energy source?

Even if you have solar panels you are still tied into the grid and cannot expect to pull electricity from them if it goes down. If any parts of the grid are impacted by a small issue, large sections can also be affected.
The expense of connecting every home and community with above or belowground wires seems cumbersome and prone to disruption.
My intuition feels that our current centralized grid system leaves us on shaky ground. I am hoping that as consumers we may have the opportunity to opt out instead for small decentralized alternatives.
Affordable single home sized windmills or grid-free solar are on my list.

decentralize the grid

2. Instead of attaching everyone to an oversized central electric grid, prone to hackers and disruptive disasters, why not create independent community systems running off alternative energies? Rural locations could purchase single home units or co-op neighborhood stations. We lose power out here from time to time and it is not a surprise. Tilted power poles loop several miles across the marsh to get here from the end of a distant road. Serving our location cannot possibly turn a profit for the electric company. Centralized power is counterintuitive and is expensive to maintain. Picture a future with wire and pole-free vistas.

Puerto Rico would have been the perfect place to experiment with such a notion after Hurricane Maria. Even if energy sources were community based, rather than countrywide, spot outages would be far easier to fix. This was a blank canvas to explore alternative infrastructure better suited to what future weather events may bring to that island community.

standard tops & bottoms

3. I have a cupboard full of container bottoms and tops with only a few that actually fit each other. If industry set standard sizes for all containers that producers manufactured, any container would be easily reusable.

responsible industry

4. Companies should be responsible for their product for its lifetime, from manufacture to how it is dealt with as waste post-use. For instance, medical supply waste or pharmaceuticals present real danger in their disposal. Some of the concerns against big wind farms are that they are not responsible for decomissioning their windmills down the line. A budget and department devoted to post-use issues should be part of every producer. This regulation might inspire them to design better for the long term.

sprawl buster

5. Big box stores and chains should be responsible for the buildings and massive paved parking lots they leave behind if they should close. A green tax for developing natural or agricultural areas could push investors to reuse exisiting buildings or sites.

**How ironic
is it that after a flood or hurricane,
people suffer from a lack of clean water?**

rainwater collection systems everywhere

6. For centuries, in water challenged areas like the Middle East, every building was built to collect and maintain its own supply of rainwater in a basement tank or cistern. Fort Pulaski, built in 1861 between Savannah, Georgia, and Tybee Island is built so the entire exterior wall rim catches and funnels rainwater into a deep cistern within the fort wall. The addition of runoff collection systems should be added to new buildings. Older buildings can be retrofitted using gutters and tanks. We catch and use rainwater for our garden as the water in our well has been compromised by salt water encroachment.

**what if every home was fitted with a system
to collect and use rainwater?**

Runoff in urban and suburban areas scoops up toxins like pesticides, fertilizers, and petro chemicals as it flows off roofs or pavement into streets and waterways. Excess urban runoff might be routed to central storage facilities. Save water and reduce runoff for two wins.

closed loop water systems

7. Municipalities must maintain intricate water systems to service everyone with clean drinking water while collecting gray water and sewage. Clean water is piped through miles of pipes, much of them old, corroded and leaching toxins like lead. Next, the system must extract and transport both gray and wastewater miles back out to be processed at a big sanitation plant.

A smaller in-house system could save loads of tax dollars and run more efficiently by reusing gray water a second time before it is directed to wastewater within a loop system. Gray water is what disappears down the drain. While 80 percent of it is nearly clean it goes out with the waste. Gray water would be looped through a filter and reused through the septic system. Collected rainwater could augment expended gray water in a final loop out to water the garden, filtering into the soil or be collected and routed to a central use facility to be sanitized for reuse.

waste not

8 . What if home waste was dyhydrated to be used in fertilizers, composting, or as home fuel? Digesters collect and dry animal manure on large farms that may be used in the production of biomass energy. In many cultures dung has been used for centuries both for heat and as cooking fuel. Human feces can also be dried if collected in a type of dry toilet, like an incinerating toilet. Big concentrations of waste are harmful to the environment, yet we funnel everyone's waste to central areas. Why not take the closed loop septic system one step further and deal with the dung on site. Dried matter can be used in fertilizers or biomass fuels.

Can we retool for a less harmful way of living? Instead of large, expensive to maintain command-centered infrastructure, could a change to smaller decentralized solutions result in less impactful, more flexible operation of community water or energy systems?

reflecting on roads

9 . We nervously watch as our glaciers melt. This process is speeding up as sunlight reflecting ice is replaced by increasing areas of heat-absorbing landmass and oceans. What if, instead of dark, heat absorbing pavement, all roads and parking lots were resurfaced in a light reflective "ice like" color? Heat reflective roads might be more durable and require less maintenance as they would expand less during high heat.

pay per mile emissions tax

10 . Why not design an EZ pass that clocks a small charge for every mile a driver travels. Drivers feel it is cheap to drive and think little about the true costs.This could assist with pro-environment infrastructure costs, help to raise carbon awareness, and encourage use of mass transit. Air travelers would also pay in to compensate for their flight milage.

a tree revolution

11 . Arbor day was started to inspire massive tree planting. We spend loads of tax dollars mowing along roads and highways. New corridors of forest would decrease this expense while absorbing CO_2 emissions and adding a buffer. Trees would beautify every street, offering shade in summer, animal habitat, and a touch of nature.

concentrated community

12 . Our current growth pattern is to flood single story retail and homes over acres, crowding out natural areas and wildlife. This outdated plan dates back to the '50s postwar love affair with cars and the dream of suburbia. It is unsustainable in today's world.

If big box stores are to rule the world for the foreseeable future, why not follow their lead. The massive store structure can be the new town square. Why not require future big box stores that displace countryside to take advantage of their building's footprint and build up. This would provide walkable living space for elderly or less mobile folks. A rooftop park, library, or gym on the top of an adjoining parking garage could add benefit to what was lost space. Big box communities could also serve urban food deserts by locating in blighted areas as the base for a pod community.

banish plastic water bottles

13 . I saw stacks of 24-case plastic water bottles for $2.25 each (cost per case!) prominently placed as you enter Big Lots. To wean the public off this devastating habit, we may need to add an eco-tax to their price. Ideally public water fountains and refillable drink containers should suffice. Bottled water is a BIG and growing business, spilling into flavored waters. A returnable deposit on containers may be another way to regulate it.

bring it for bulk

14 . At a party, I ended up in an interesting conversation that tested all the guidance for civility offered in this book's first chapter. I was speaking with a fellow who is in charge of building new dollar stores in our tristate area. It seems as if they are popping up everywhere—even at the turn off to our road in what had always been a very rural area. The gist of our exchange is that flooding the stores everywhere was economically a numbers game that sometimes left behind empty buildings and parking lots. Dollar stores are big contributors to our plastic waste problem, offering cheaply made, single serving, heavily packaged products for a few dollars. While each of us may have spoken from opposite sides of the political spectrum, I believe we both came away thinking beyond our initial construct. While I am troubled by much of the product content and single story sprawling construction of these stores, their numbers may represent an opportunity. My husband suggested these chain stores may represent a modern version of the traditional general store. Dollar General Store's name is based on this concept, serving rural or pocket neighborhoods a wide variety of basics along with a few healthy offerings like milk and frozen broccoli. This could be widened using their local presence to offer customers many more environmentally sound alternatives.

What if instead of shelves of unhealthy snack type fare, a section was replaced with unprocessed bulk foods like dried fruit, cereals, beans, flour, rice, nuts, and the like. People could fill their own containers, saving money and omitting packaging from both the manufacturing and refuse ends of delivering food to consumer. Even products like dishwashing liquid or pet food could be collected in reusable containers. This healthy coop style food distribution commonly found in upscale communities would reach a demographic usually offered only low nutrition, high sugar and salt products responsible for a nationwide epidemic of obesity and diabetes. Dollar stores could reap higher profits while gaining positive press as a green retailer becoming part of the solution.

I will look forward to speaking further with my friend at the next party :)

* *One of the thinkers, Heidi Thompson, was inspired to puzzle out a design for a hygienic bulk container that dispensed and self weighed apportioned product so any container could be filled by customers.*

The more minds thinking creatively the better!

donation days

15 . If you have ever been through a college town on the day after graduation, it is hard not to be horrified by the piles of mattresses, furniture, lamps, and other dorm room debris overflowing dumpsters and lining the streets surrounding campus. A good lesson to learn has clearly been missed here. Just as companies should be mindful of the post-use disposal of their products, colleges should address a better solution to short-term use of their residences and community housing. Why not introduce clear practices for students (and people in the process of moving) for how to sort and where to place the things they have used and no longer want to take with them. Give the landfill a break and reroute these materials to organizations, like the Habitat ReStores, area thrift stores or recycling venues.

Schedule a day for all good furniture to be delivered to a central location for local organizations to pickup. Similarly, define a location for carpets, working or broken electronics, extra food, clothing, or miscellaneous items. These days should enjoy the same amount of preparation and participation as graduation. The speech accompanying them might be on how to be responsible as an adult.

designing for demise

16 . Apple products are quickly outdated and notoriously hard to repair, reflecting their mantra drumbeat of "buy our new version now." Airpods are a good example of the company's laser focus on design over long-term use or ease of recycling. Airpods cannot be recycled as their lithium ion batteries can catch fire. They actually must pay recyclers to deal with these tiny headphones, because it is not profitable to employ someone to wrestle the tiny battery and precious elements out of its sleek design. A final requirement for any product designer should be easy removal post-use of any recyclable, dangerous, or valuable components, especially if they are working for a company hell-bent on continually shoving their latest model into our hands. Re-instituting a returnable deposit on all reusable material, including plastics, glass, metal, and other recyclables, would will raise awareness, reducing waste and litter.

Products are made of valuable materials and should always be regarded as that, rather than as garbage.

come together

17 . Why not institute a United Nations based World Environmental Coalition to guard aginst problem regimes and bad practices. Brazil's far-right president Jair Bolsonaro has swept away important protections for the Amazon rain forest. Every minute, a football sized section is destroyed for logging and large scale cattle ranching. To stop alarming actions like this, tarifs or embargos could be imposed. The vast Amazon is one of the largest reservoirs of sequestered carbon dioxide, released as the forest is cut. The living trees there act as the lungs of our planet.

the can-do soft-shoe

It is the dreamer that looks beyond a wall. Big companies will continue to operate as they have because bigger has been profitable for them. But interesting advances in biotech design based on nature's systems may offer consumers alternative earth friendly products. Researchers at Cornell University are employing (tree-inspired) capillary action rather than pumps to move water, using the walls in a building.
The 2015 Paris Climate Accord threw a gauntlet to scientists and biotechnology firms to accelerate their efforts in creating technologies that would help the world to keep atmospheric CO_2 levels in check while preserving the environment.

New technologies are in development for solar energy conversion, generating electricity, and synthesizing carbohydrates from atmospheric CO2 for fuel and as raw material for the chemical industry.

Plant biotechnology to increase food production and carbon sequestration.

Mitigate greenhouse gas production to preserve the environment and biodiversity. -*Embo reports 2016*

When faced with adversity, people have risen quickly to the challenge, working to find good solutions. We may not be able to convince those around us with words that now is the time to focus on earth's future instead of all the shortsighted daily diversions that distract us. That is why real strength is in engaging yourself to take action and set an example.

If you want to change the world,
start with yourself.

. . . . Mahatma Gandhi

the big picture

..........developing a long view & deep vision

tools .

recipes for success

the long view

a star to steer by

the journey is the thing

Just as a reputation is built over years, retooling how you think and operate is a practice built on personal integrity. After 30 years in our community, it's clear that our actions define who we are. A focus on being the best person you can be enriches the character of who you are.

Similarly, each action you take in your solo battle to restore health to our planet is a block of gold. Changing old habits to new is made easier by understanding the big picture. Things we have done or accepted have created many of the present challenges we face now. In the big picture there are two important things.

You, which is to say all of us, acting for the long view goals of retooling our daily actions for a healthier future. The second and more important thing than us is nature, sustaining all life on earth. Instead of considering that nature is here for us to exploit, the big picture is clear: we must work for the good of nature.

DIY or BUY

Many of the ingredients in products we routinely turn to are borderline frightening, bad news for the environment, and may even be unhealthy to use. I will admit that, for the most part, I haven't been paying much attention to the herd of plastic containers under the sink and lurking in the bathroom. After getting schooled by reading piles of articles on the industrial practices of our current chemical age, I began to search for healthier alternatives. The good news is there are plenty of green choices on the market. You can use the following section to cull your cupboards for a healthier home or take a further step to make your own cleaners and personal care products. The recipes that follow are easy to make, work well, leave more space in my cupboards, and money in my pocket.

recipes for success

If you need it, you can make it.
Nearly everything you need
is already in your kitchen.
Here is where we go old school,
back when things were simple.

A few years ago a generous friend of mine gave me a set of tarnished copper pans. She didn't want to deal with them; they were too much work to polish. I imagined I would get them gleaming, but then the sad pans sat on a shelf in my house gathering dust for several years.

In researching this book, I have learned an astounding amount. One of the most enlightening things I stumbled upon is that with a few ingredients I can skip having loads of unknown chemicals in plastic containers crowding my cupboards. "Clean copper with lemon juice," I found in a 1950's guide to housekeeping. If that works why had I paid four dollars for special copper cleaner? Because I believed the marketing apparently.

Spoiler alert: Lemon juice works beautifully!

I really enjoyed polishing those copper pans with lemons. It's kind of a sparkly zen meditation. Next, I tackled our shower with baking soda & vinegar. This experiment did require some elbow grease, but came out equally clean to the section I scrubbed with Comet and a few other brand cleaners. There is something very satisfying about using simple known ingredients to accomplish the same effect you thought required heavily produced commercial cleaning products.

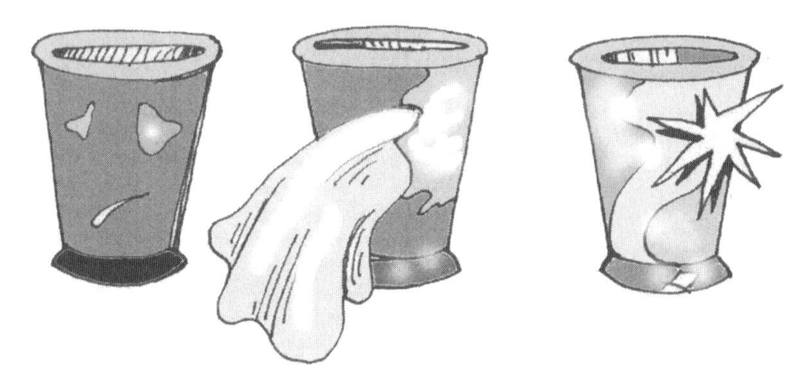

Polishing is a lost art in this busy world. My friend Drew grew infinitely happy anytime he found an old blackened silver cup at the flea market. He would rush home to polish it, delighted to add it to his large collection of gleaming silver cups to drink rum from.

Here is some vintage info from a housekeeping manual written over 90 years ago:

. easy to make polishes .

copper . lemons doll up copper!

pewter . make a paste of flour, salt & vinegar

silver . lemon & salt

chrome . vinegar & baby oil

NEWS UPDATE! I spilled some tomato sauce on my copper sauce pan, which made it shine tenfold, leading me to smear it with ketchup. Wow! Wait a few minutes, gentle buff & rinse. This works even better than lemons.

By the way . . . because many people are not interested in polishing silver and other metals, if you are so inclined, dash over to the thrift store and rescue some pieces for mere pocket change. Back in the day, these were highly valued items. Wrap your polished silver in flannel and store in a cool dark place. It is fun to pull a few posh things out to perk up a party.

Truth is ever to be found
in simplicity,
and not in the multiplicity
and confusion of things.
..... Isaac Newton

Think of the thousands of cleaning products marketed to us. Unlike food products, there is no law requiring cleaning product manufacturers to list all the ingredients used in their products. We bring toxic whatnot into our homes, exposing our families and pets to all sorts of unknown synthetic chemicals. *The U.S. National Center for Health Statistics states that one in three people suffer from allergies, asthma, sinusitis, or bronchitis. Some cleaning chemicals are allergy or asthma triggers. Many contain ingredients harmful to our lungs. (info source **EWG Environmental Working Group**)

The average American household contains around 62 toxic chemicals, from the noxious fumes found in oven cleaners to the dangerous phthalates in synthetic fragrances. Routine exposure over time has been linked to asthma, cancer, reproductive disorders, hormone disruption, and neurotoxicity. Some chemicals build up in your body through exposure, others can trigger serious reactions or diseases. Some 40 percent of the chemicals from our personal care and household cleaners end up in the air affecting the ozone layer or in our water. But pollution is not just in our water and air, it also is absorbed into us.

Research into the health effects from the normal use of cleaning products found they can be damaging to our respiratory systems. These chemicals irritated the delicate mucous lining of the lungs, and can causie permanent damage resulting in an overall decline in lung health. Gaining a clear understanding of what we are exposing ourselves, our families, and our environment to in common household products is eye opening.

If you want to continue using commercial products limit the quantity you use and buy fragrance free.

The most dangerous chemicals currently used in home cleaning products

Dioxane . Suspected carcinogen . common detergents.

Quaternary Ammonium Compounds or "Quats"
Asthma triggers . spray cleaners and fabric softeners

Chlorine Bleach . Bleach fumes contain chlorine and chloroform linked to respiratory or neurological effects and cancer. Bleach forms dangerous gases when it reacts with ammonia or acids like vinegar. Bleach also kills the helpful biome in septic tanks that work to break down solids.

Formaldehyde . A preservative and known carcinogen.

Ammonia . Respiratory and skin irritant.

Perchloroethylene ("PERC") . In spot removers, home dry cleaning products, and upholstery cleaners . A carcinogen and neurotoxin.

Antibacterials . In 2016 the FDA banned triclosan along with 18 other antibacterial compounds from hand & body soap, but these may still be found in cleaners. Creates endocrine disruption and antibiotic resistance.

2-Butoxyethanol (also 2-BE, BCEE, or Butyl cellosolve)
Laundry stain removers, oven cleaners, degreasers
Skin and eye irritants. These are listed as toxic substances in the Canadian Environmental Protection Act.

Diethylene Glycol Monomethyl Ether (also DEGME or Methoxydiglycol)
A solvent in some degreasers and heavy-duty cleaners. (Banned in the EU.) Linked to reproductive health issues.

Fragrance

A common ingredient listed as "fragrance" contains hundreds of different chemical compounds, including phthalates, an endocrine disruptor.
Fragrances may also trigger asthma and allergies.

**Many common cleaning products will burn or irritate your skin and eyes. They are fatal if swallowed. After we use these chemicals, they may wash into our water table or be disposed of in landfills leaching out over time.

A list to NOT shop for. . . . .

Air fresheners, fabric softeners, and dryer sheets

Contain chemicals known to cause asthma in otherwise healthy people.

Any scented cleaning product

Remember fragrance is a loophole ingredient for other chemicals

Antibacterial soaps and products

Can encourage the development of drug-resistant superbugs.

Corrosive drain cleaners, oven cleaners and toilet bowl cleaners

Contact can cause severe burns on the skin and eyes, irritation to the throat and esophagus. These will find their way into water systems.

Bleach and ammonia

Each produces fumes with high acute toxicity to eyes, nose, throat, and lungs. If combined they create a toxic gas that can cause serious lung damage.

Suds . like shampoo, liquid soap, bubble bath, laundry detergent

1,4-dioxane, diethanolamine (DEA), triethanolamine (TEA), sodium laureth sulfate, PEG compounds, etc. are known carcinogens linked to organ toxicity.

The real deal cleaners

DIY. If you make it yourself, you will save money and know your ingredients. Reuse clean spray bottles and other containers over and over. Many companies now offer non-toxic products, but they still come in new containers, and will need to travel to get to you.

Why not keep it all in-house?

Baking Soda

Cleans, deodorizes, softens water, and scours.

Soap

Unscented liquid, soap flakes, powders, or bars are biodegradable. Castile soap is best (like Dr. Bronner's). Avoid soaps that contain petroleum distillates.

Lemon Juice

Super strong food acid effective against most household bacteria. Reuse squeezed lemon to wipe down cutting boards.

White Vinegar

Use white vinegar to cut grease; remove mildew, odors, and some stains. Can prevent or remove wax buildup.

Washing Soda

Washing soda or SAL Soda is a mineral (sodium carbonate). It cuts grease; removes stains, softens water, and cleans walls, tiles, sinks, and tubs. Use with care, since washing soda can irritate mucous membranes. Do not use on aluminum.

Vegetable or Olive Oil

Wood polish. Add lemon to oil on unvarnished wood.

Cornstarch

Clean windows, polish furniture,
shampoo carpets and rugs to remove stains.

Use these nontoxic alternatives, please.

Alcohol

Disinfectant. You can use other forms of ethanol: vodka is a potent odor remover and also nice in a Bloody Mary.

Hydrogen Peroxide

Disinfectant for wounds or surfaces in the kitchen or bathroom. It also will remove stains in fabrics or tile grout.

. cleaners .

borax & lemons in the laundry

baking soda & vinegar in the shower

water & rubbing alcohol to clean marble or granite

lemon & salt to clean rust

tree oil antibacterial and antimicrobial properties
add 2 tsp to a spray bottle of water for an all-purpose cleaner

white vinegar & water to rinse clean glasses
*For sparkling clean windows:
Mix 2 cups of hot water
1/4 cup of vinegar
1 tablespoon of cornstarch.
Mix very well & pour into a spray bottle.
Use with crumpled up newspaper.

* Simple fixes *

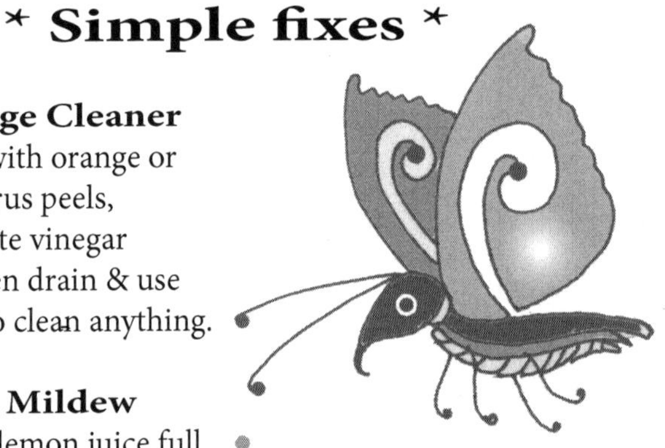

April's Orange Cleaner
Pack a Ball jar with orange or
various citrus peels,
fill with white vinegar
mull a week, then drain & use
in a spray bottle to clean anything.

Mold and Mildew
White vinegar or lemon juice full strength. Apply with a scrubby.

Moth Deterrents
Mothballs are made of ***paradichlorobenzene,*** which is harmful to the liver and kidneys. Use cedar oil in an absorbent cloth or make a satchet of dried lavender, rosemary, mint, rose petals & lemon peel instead. Lovely!

Toilet Bowl Cleaner
¼ cup baking soda &
1 cup vinegar
Pour into bowl for a few minutes
Scrub with brush and flush.

Tub and Tile Cleaner
Rub in baking soda with a damp sponge and rinse with fresh water. For tougher jobs, wipe surfaces with vinegar first and follow with baking soda as a scouring powder. (Vinegar can break down tile grout, so use sparingly.)

Refrigerator Cleaner
Use an all-purpose spray of vinegar and water for cleaning the interior of refrigerators and the outside of refrigerators made of materials other than stainless steel. To clean the outside of a stainless steel refrigerator, use undiluted vinegar to avoid streaks.

Water Rings on Wood
Water rings on a wooden table or furniture are the result of moisture that is trapped under the topcoat, but not the finish. Try applying toothpaste or mayonnaise to a damp cloth and rub in.

Furniture polish
1 part white vinegar
to 3 parts olive oil
Use clean recycled rag
to wipe onto wood.

DIY dishwashing soap

1 ¾ cups boiling water
1 Tbsp borax
1 Tbsp grated bar soap (use homemade soap, castile bar soap, Ivory)
15–20 drops essential oils, optional
(Oils such as lemon, eucalyptus, sweet orange, geranium, or lavender have antibacterial properties.)
Mix lemon & eucalyptus oil to cut grease.

Heat water to a boil.
In a medium bowl, combine borax & grated bar soap. Pour boiling water in, whisking until the grated soap is completely melted.
Mixture should cool for 6–8 hours.
Stir occasionally. Dish soap will begin to gel. Pour into a dispenser, add essential oils (if using). Shake well.

Note:

Commercial dish soaps add harmful chemical surfactants to create suds (sodium lauryl sulfate). Suds do not mean it cleans better. This dish soap will cloud the water with few suds but cleans your dishes very well.

Block pests without using poison or toxic chemicals. Pesticides are linked to the onset of diseases like Parkinson's, Alzheimer's, brain cancer, infertility, and birth defects. Applications can spread by wind or runoff, contaminating groundwater, and placing delicate ecosystems in danger.

flies & moths . keep live basil plants in your kitchen, or dried bouquets of basil, rue, mint, tansy, bay leaves, or pennyroyal in the cupboard to repel moths. Make an easy fly trap. Fill a jar with sugar water and roll a paper funnel. Put the funnel, wide side up, in the top of the jar. The flies fly into the jar for dessert and never come out.

spiders . dislike citrus. Spray a mix of water & lemon juice around doorways and window sills. Deter spiders in your garden by spreading around lemon, orange, or lime peels.

ants . use a 50/50 solution of vinegar/water to wipe down anywhere you've spotted ants to destroy the scent trails they use to navigate.

mosquitoes . catnip repels more effectively than DEET. Grow it in your garden or apply undiluted catnip oil to the skin for up to two hours of protection. It will also make your cats very happy.

fleas . salting and vacuuming your floors will kill flea eggs. But since fleas have a three-day reproduction cycle, salt every day for nine days and vacuum every third day. Empty the vacuum each time, or the live ones will jump back out! Crush garlic, adding that and brewer's yeast to your pet's diet. Rub tea tree oil underneath your dog's collar.

fruit flies . fill a glass three-quarters full with cider vinegar, add seven drops of dishwashing liquid, and top off with warm water.

cockroaches . mix three parts boric acid and one part powdered sugar. Sprinkle it under and behind the refrigerator, stove, dishwasher, under the sink, and into cracks along the edges of cabinets. Keep your kitchen tidy.

mice . engage a cat or put peppermint oil on cotton balls. Tuck anywhere mice might go. Wash surfaces with vinegar & peppermint oil.

DIY bug repellent

1 cup coconut oil
½ cup shea butter
½ cup beeswax
¼ cup fresh rosemary leaves
1 tsp dried whole cloves
2 tbsp dried thyme
½ tsp cinnamon powder
¼ cup dried catnip leaf
2 tbsp dried mint leaf
Vitamin E oil
Essential oils: lavender and lemon (10+ drops of each)
Ice cube tray to use as a mold

Use a double boiler to heat about 1 inch of water (in bottom pan) until just starting to boil. Place the dried rosemary, cloves, thyme, cinnamon, catnip, and mint in the top part of the double boiler adding the coconut oil. Cover the double boiler keeping the water at a low boil for 30 minutes or until you can smell the rosemary and the oil has become a darker color.

To make a stronger infused oil fill a crock pot about half full with water, put the coconut oil and herbs in a glass mason jar with a tight lid, cover, and keep on lowest setting for several days.

Strain the dried herbs out of the oil using a small mesh strainer or cheesecloth and pour the oil back into the double boiler. The oil will probably be reduced by almost half and you should have about 1/2 cup of infused oil. If you have more, save it—use just the 1/2 cup.

Add the shea butter and beeswax to the infused oil and stir until all are melted. Remove from heat, add Vitamin E oil and essential oils. Pour into molds.

Leave in the molds until completely set (overnight is best) or you can speed up the process by placing in the fridge. Store in fridge or under 80 degrees.

Rub the bars on exposed areas of dry skin to protect against mosquitoes. It's a wonderful skin lotion also.

**Science shows us truth and beauty,
filling each day with a fresh wonder
of the exquisite order which governs our world.**

. . . . Polykarp Kusch

Commercial beauty products are full of harmful chemicals, many of which have not been tested for safety in humans. Others are tested on animals, subjecting them to torturous circumstances. This is another big industry based on you purchasing their product in plastic containers over and over. Most shampoos actually strip your hair of natural oils forcing you to follow them with a conditioner or other products. Then 89 percent of these chemical concoctions just wash down the drain into our water supply. Keep these simple ingredients that have been used for centuries on hand instead. They are good for you and other living things.

Coconut oil, shea butter, beeswax, carrier oils like olive or almond, essential oils like lavender, mint, or lemon, arrowroot, baking soda, honey, salt, dried herbs like chamomile & calendula, and liquid castile soap.

. natural deodorant .

Note: Your body was built to sweat and naturally excrete toxins. The marketing of deodorants began in 1880 aimed at young women worried that their social standing would suffer if they were discovered to have stinky underarms. Today, Americans spend 18 billion dollars a year to purchase a toxic cocktail of chemicals, hormone disruptors, carcinogens, and even pesticides. Endocrine-disrupting chemicals such as parabens are known to produce adverse developmental, reproductive, neurological, and immune effects. They can affect your body's growth, metabolism, tissues, function, mood, and ability to sleep. Carcinogenic aluminum compounds such as aluminum chlorohydrate used for the purpose of blocking the sweat ducts have been linked to breast cancer and Alzheimer"s. Triclosan, a registered EPA pesticide that kills bacteria, fungus, and other microbes, is regularly used in deodorants, as well as some soaps and other personal care products.

. diy deodorant .

¾ cup arrowroot powder/non-GMO cornstarch
¼ cup baking soda
4-6 tbsp melted coconut oil
optional . 10 drops total . lavender and/or bergamot essential oils

Mix baking soda & arrowroot powder/cornstarch, stirring in melted coconut oil and essential oils, mixing with a fork into a smooth paste. Transfer mixture to a jar with a tight fitting lid or reuse a roll-on dispenser.

. shampoo .

Note: the word sham before poo. Most people can skip using it. The singer Adele (who has gorgeous hair) hasn't washed hers in years. I stopped many months ago, and frequently hear: "Your hair looks so nice." I just rinse my hair in water and use my DIY conditioner. But if you adore shampooing give this a try.

. shampoo .

¼ cup coconut milk (homemade or canned)
¼ cup liquid castile soap like Dr. Bronner's
20 drops of essential oils (peppermint, lavender, rosemary, or orange)
For dry hair: add ½ tsp olive or almond oil (optional)
Rinse: A 50/50 mix of apple cider vinegar/water rinse helps increase shine.

. conditioner .

3 tablespoons coconut milk (rich in essential vitamins)
1 tablespoon extra-virgin coconut oil (moisturizes hair)
15 drops lavender essential oil (antibacterial, boosts hair growth)
3 tablespoons distilled water (tap water can compromise mix)
Whisk well by hand or with mixer. Place in reused pump container. (Lasts about a month, stored in dark place.) If you use this without shampooing put on barely wet hair for about 3 minutes, then rinse out (using the 50/50 vinegar rinse).
OR make a conditioner of 1 part lemon juice to 4 parts aloe vera gel to stimulate hair growth, add shine, and help restore the pH balance of the hair.

. intensive hair conditioner .

Blend avocado & coconut milk in a blender and apply to dry hair.
Leave on up to 30 minutes and rinse.
There are loads of variations including adding an egg (shine), honey (keeps hair hydrated), banana (helps frizzies), olive oil (stronger hair) or plain yogurt (adds protein to hair & cleanses scalp).

raw organic or manuka honey . face mask & cleanser

Honey promotes natural antibacterial healing through enzymes and probiotics that gently nourish skin.

Mask . Open pores with warm water. Rub a quarter-sized dab of honey in your hands, working in a circular motion to massage into skin. Leave on for 10 minutes. Wash off with warm water, patting dry for soft, balanced skin!

Or add . Vitamin C powder, 2 drops of organic lavender essential oil.

Cinnamon: add a pinch of powder to honey in palm for oily skin.

Lemon: mix a tsp with honey in hand to reduce age spots, tighten pores.

. natural sunscreen .

Sunshine is good for you, producing vitamin D, lifting your mood, and providing many other health benefits. Sunburning is the issue.

Note . Recently there has been a big push to make sure everyone has slathered on sunscreen for any period of time in the sun. Sunscreen, which is composed of many chemicals designed to block intake of healthful sunshine, also has "penetration enhancers" that help the product adhere to skin. Many absorbed sunscreen chemicals can be measured in blood, breast milk, and urine samples. The best sunscreens are proactive measures such as: limiting sun exposure by working in shade, early or late in the day, and wearing hats or other protective clothing. Eat a diet rich in saturated fats and monounsaturated fats, omega–3s (fish), antioxidants and lots of leafy greens while avoiding processed foods, vegetable oil, grains, and sugar. A daily dose of 2+ tablespoons of tomato paste provides lycopene to help skin protection.
Boost your intake of vitamins C and D during heightened sun exposure.

. DIY . sunscreen lotion .

1 cup coconut oil
1 cup total of any or mix of shea butter, cocoa butter, or mango butter
1 cup beeswax
2 tablespoons (or more) of zinc oxide
optional: 1 tsp vitamin E oil, lavendar essential oil (not citrus)

Place coconut oil and butter (not zinc oxide) in a double boiler & bring saucepan water to a boil stirring oils until melted smooth. Remove from heat to add the zinc oxide powder, vitamin E, and any essential oils, gently stirring by hand.
Pour into mold. (ice cube tray works for small blocks.)
Keep in fridge or freezer to feel really cool when you use them.

. tooth care .

Healthy teeth reflect overall well being

Note . In the 1930s, some traditional societies whose diets were rich in fat-soluble vitamins A, D, E and vitamin K2 were discovered to be free of degenerative illness and cavities. Glycerin keeps commercial toothpaste smooth but hinders natural remineralization. (Calcium, phosphate, & other teeth-rebuilding elements contained in saliva attach directly to and harden enamel preventing tooth decay.) Brush gently in a circular motion around each tooth after meals, floss (using plastic-free floss) & rinse.

. make toothpaste .

½ cup filtered water . keep it pure
¼ cup Bentonite clay . absorbs toxins, heavy metals, & impurities
2 tbsp Calcium/Magnesium Powder . whitener & good source of minerals
3 tbsp coconut oil . natural anti-bacterial & anti-fungal binder material
¼ tsp unprocessed salt
¼ tsp baking soda . removes stains. Healthful minerals & is alkaline.
10 drops trace minerals . extra boost of good stuff. (optional)
essential oils . 10 drops peppermint oil (enhances circulation to gums and teeth.)
optional 3 drops each (bergamot, wild orange, clove, cinnamon, eucalyptus, rosemary) supports healthy immune function in the mouth.
stevia . optional sweetener
arrow root . just a teaspoon or less as a thickener if needed.

Mix all ingredients in a food processor until smooth. Store in a small glass jar.

Both the toothpaste and mouthwash work great. I never get morning mouth!

. DIY mouthwash .

2 tsp baking soda . whitens, restores balance
½ cup filtered or distilled water . free of chemicals
2 drops tea tree essential oil . antibacterial agent
4 drops peppermint essential oil . Freshens breath
2 drops bergamot oil . promotes oral health

Shake your homemade mouthwash before each use. Swish about 2–3 teaspoons in your mouth for a minute or two. Avoid swallowing it while gargling.

Note . Listerine was developed in the 19th century by Joseph Lister as a surgical antiseptic, but by the 1920s, it was marketed as a cure for bad breath, as a floor cleaner, and dandruff remedy. Mouthwash use can dry out and imbalance the pH in the mouth, killing useful bacteria that strengthens teeth.
Studies have linked it to oral cancers and diabetes.

Water, air, and cleanness are the chief articles in my pharmacy.

. . . . Napoleon Bonaparte

There are many natural remedies used successfully over centuries. They can help ease uncomfortable symptoms without a trip to the doctor or turning to commercial pharmaceuticals. Here are a few things to keep on hand. Use nature's elements to cure; it is better for you and the planet.

. first aid kit .

activated charcoal . for ingestion of toxins, food poisoning, intestinal illness, vomiting, diarrhea. Keep your local poison control number handy.

aloe vera plant or gel . heals cuts, burns, blisters.

organic apple cider vinegar . with "the mother" for digestive troubles, indigestion, food poisoning, or bolstering your immune system.
1 teaspoon per half glass of water each hour will speed your recovery.

baking soda . ¼ tsp mixed in water will alleviate severe heartburn or urinary tract infections. A paste poultice can sooth stings or spider bites.

cayenne powder . adds zip to foods and body systems. Topically, cayenne powder will help stop bleeding.

chamomile . soothes fussy toddlers, settles the tummy, reduces gas, lessensmenstrual cramps, relieves headache, calms frayed nerves, and improves sleep. It is also great for your skin.

comfrey . a dried herb that promotes healing from injuries. A leafy plantain/comfrey poultice placed on a wound speeds its healing time and prevents/reverses infection. Use it on bug bites, cuts, bruises, & poison ivy.

coconut oil . skin salve, diaper cream, antifungal agent, soothes dry skin and chapped lips. Taking 2 tbsp daily of coconut oil is recommended to protect against Alzheimer's and nourish the thyroid.

peppermint . highly effective digestive aid and nausea remedy. Applying essential oil behind the ears and on the bottom of feet helps to alleviate headache or nausea. Peppermint tea soothes tummy.

(More on essential oils pages 130-133)

slippery elm . laryngitis, sore or irritated throat. The herb can be used in teas for sore throat relief.

tiger balm . apply to temples to ease headache, near nostrils and over chest to open lungs and sinuses, on throat glands to soothe throat.

eucalyptus . use for respiratory issues. To ease sinus congestion place eucalyptus oil or herb in a face steamer. To help open nasal passageways make a eucalyptus essential oil vapor rub: dilute with coconut or olive oil to apply externally to the feet or chest.

ginger . great for nausea, reflux, stomach trouble, morning sickness, or motion sickness. Its anti-inflammatory properties can ease a dry or asthmatic cough. Use fresh to make a tea or purchase in capsules.

hydrogen peroxide . disinfectant for family and pets. Clean out wounds. To prevent ear infection put a dropper full of hydrogen peroxide in the ear. Leave for 15 minutes until it stops bubbling. Repeat with other ear.

vitamin C . always helpful for any illness, powder can be mixed into tea, drinks, or food. Naturally found in fruits and orange juice.

supplies . butterfly bandages, gauze, bulb syringe (for flushing wounds), hot water bottle, ice packs, enema kit, superglue for moderate cuts. adhesive tape, elastic wrap bandages, rubber tourniquet, cotton balls and swabs, duct tape, scissors and tweezers, hand sanitizer, antibiotic ointment, antiseptic solution/ towelettes, eyewash solution, thermometer, sterile saline for irrigation/flushing, breathing barrier (surgical mask), syringe, medicine cup or spoon, doctor's and emergency numbers, first-aid manual, hydrogen peroxide to disinfect.

essential oils

Essential oils are concentrated compounds extracted from plants (capturing their essence, scent and flavor). Purchase only pure high-quality oils extracted by distillation or cold pressing, avoiding oils that have been diluted with synthetic fragrances, chemicals or oils.

Essential oils are used in the practice of aromatherapy.
They are either inhaled, or diluted with a carrier oil and rubbed on the skin. The aromas from essential oils stimulate areas of your limbic system, a part of the brain affecting memory, emotions, and sense of smell. Many people turn to essential oils for a safe and cost-effective natural path to maintaining or regaining their health.

Always test a little amount on a small area of your skin to see how your skin responds. Wash off gently if you feel any discomfort or see a reaction. Be sure you understand how to use each oil. Some oils may be ingested but may irritate your skin. It is wise to consult with your doctor if you are taking any prescriptions. The oils may change how your body absorbs drugs or trigger an allergic reaction.

Undiluted oils should be diluted, usually with vegetable oils, creams or butters like cocoa or shea. Try about six drops in one tablespoon of carrier oil. Don't use oils on injured or inflamed skin. Older people or young children may be more sensitive to essential oils so dilute accordingly. Store your essential oils in a cool, dark place out of reach of children. Safely dispose of oils over 3 years old. Spoiled oils can irritate your skin.

Used the right way, essential oils can help you feel better naturally. Examples: Breathe in ginger vapors to feel less nauseated after chemotherapy cancer treatment. Tea tree oil fights bacterial or fungal infections, including MRSA bacteria.

There are a wide variety of essential oils. I have listed just a few and some of the qualities each offers. Please further research how to correctly apply them and the quantity to use.

I start my day with a tall glass of water with one drop each of lemon and sweet orange essential oils. It is cleansing and tastes like fresh potential. A sniff of the sweet orange is joyful.

lavender oil . eases headache, improves sleep and offers stress relief. It is a sedative, antispasmodic, anti-anxiety, anti-inflammatory, antimicrobial, antioxidant, antibacterial, anesthetic, immune-boosting, and antiviral.

rosemary oil . is a stimulant. The aroma of rosemary increases heart rate, blood pressure, and respiratory rate boosting your immune system. Rosemary infused cream will increase brain wave activity while decreasing levels of the stress hormone cortisol.

sweet orange oil . inhaling the aroma reduces anxiety. When applied topically, a calming effect slows pulse, while inducing a more cheerful and vigorous feeling. It's delightful!

lemon: used to aid digestion, mood issues, and headaches. It is also a powerful antibacterial, astringent, and antiseptic agent. Diluted in creams it lightens age spots, diminishes the appearance of wrinkles, promotes circulation, and tones the skin

*** Citrus oils that are safe in your food may be bad for your skin, especially if you go out into the sun. Use in products that can be showered off.*

tea tree . used to fight infections and boost immunity, to treat coughs and colds, heal wounds, and alleviate sore throats, skin ailments or acne. It kills oral bacteria like gingivitis for up to two weeks used in mouthwash.

cinnamon oil . can be used to help clear up chest colds. Applied topically, it will soothe muscle aches and pains. Contains powerful antioxidants.

lemongrass oil . treats fevers and gastrointestinal issues. It is a potent insect repellent. Its antifungal properties will combat a nasty yeast associated with dandruff. There is evidence that it has the potential to slow the growth of cancerous cells and tumors. Add to bath if you feel you are getting sick.

eucalyptus oil . diffuse it at home to freshen up a room. It also makes a great pantry moth and bug repellent. As a pesticide it will kill fungus, bacteria, insects, mites, and weeds. Never ingest.

more essential oils . naturally useful

clary sage oil . regulates hormones, diluted in a massage oil with lavender and marjoram makes an effective treatment for alleviating menstrual pain and cramping.

frankincense oil . an immune-enhancer capable of destroying dangerous bacteria, viruses, and even cancers.

sandalwood oil . used to calm nerves and help with mental focus. Helps to reduce the proliferation of skin cancer (topical application containing five percent sandalwood oil). Reduces wrinkles.

bergamot . reduces stress and depression, improves skin conditions like eczema. Positive effects on your cardiovascular, digestive, and respiratory systems by lowering blood pressure, pulse, stress and anxiety levels, and cortisol levels. It is a natural deodorant and boosts oral health.

peppermint oil . used to boost energy, help with digestion, soothe nausea, improve concentration and memory. It is an analgesic (it numbs and kills pain on the skin). Inhaling the aroma can ease a tension headache. Its fresh scent and antimicrobial, antifungal, and antioxidant properties are great additions to DIY toothpaste, mouthwash, or lip balm.
.... A 2013 Journal of the International Society of Sports Nutrition article found that peppermint oil increased brain oxygen concentration, improved exercise performance, and reduced exhaustion in healthy male athletes who consumed peppermint oil with water for 10 days.
.... Improves brain function, neuroprotective effects, cognitive performance boosting abilities, preventative for Alzheimer's and dementia.

There are over one hundred essential oils, each with many uses.
If you want to just get a few to try, the top 5 are:

peppermint, tea tree, lavender, lemon, cinnamon

You can find more information
about each essential oil online along with
easy recipes for DIY home, health, and personal beauty products.

Better health for you & the environment

energy booster

peppermint, grapefruit, lemon, lemongrass, eucalyptus, and rosemary.

mood booster

antidepressant

lemon, clary sage, lavender, rosemary, sweet orange, roman chamomile, bergamot, ylang ylang, rose, frankincense, jasmine, and vetiver.

immunity booster

oregano, myrrh, ginger, lemon, eucalyptus, frankincense, peppermint, and cinnamon.

powerful antibacterial

astringent & antiseptic

lemon, cinnamon, lemongrass, clary sage, lavender, tea tree, and sandalwood.

improves skin & hair

lavender, roman chamomile, frankincense, tea tree, geranium, myrrh, helichrysum, rosemary, and clary sage.

digestion issues

peppermint, fennel, lemongrass, marjoram, black pepper, and juniper berry.

helpful to women

balances out estrogen and progesterone levels in your body, helps infertility, PCOS, PMS, and menopause symptoms

clary sage, geranium, and thyme.

reduce toxicity

internal detoxification

lemon, grapefruit, parsley, fennel, lemongrass, peppermint, and ginger

reducing toxicity in your home or work

grapefruit, orange, lemon, lemongrass, eucalyptus, cinnamon, peppermint, and tea tree.

aches and pains

lavender, peppermint, eucalyptus, chamomile, rosemary, marjoram, thyme, frankincense, turmeric, ginger, and myrrh.

In a 2014 study of 60 neck pain sufferers, a cream composed of marjoram, black pepper, lavender, & peppermint oils was applied for 4 weeks after bathing. Improved pain tolerance was shown in the neck as well as significant improvement in the 10 motion areas that were measured.

J Altern Complement Med. 2014

Coughs. help loosen phlegm, clearing irritants and infections from the body. To soothe coughs try brewing up a soothing ginger tea by adding 1tsp grated fresh ginger, honey, and lemon juice to a cup of hot water. Steep for a few minutes before drinking. Add a few drops of eucalyptus oil into a diffuser or steam bath—or inhale aroma to clear airways.

Hydrate a cold or flu. drinking liquids at room temperature can alleviate a cough, runny nose, and sneezing. Hot beverages ease a sore throat, chills, and fatigue (clear broths, herbal teas, decaffeinated black tea, warm water, and warm fruit juices). Taking 3 drops under the tongue of frankincense & oregano essential oils 3 times daily for a week will also help relieve a cold.

Steam. will help a wet cough (one that produces mucus or phlegm). Take a very hot shower or bath, allowing the bathroom to fill with steam. Stay in this steam for a few minutes until symptoms subside. Bundle up and get in bed to sweat out the fever. Have a long deep sleep. Drink plenty of water afterward to cool down and prevent dehydration. (This was my dad's immediate response to any illness.)

Saltwater gargle. one of the most effective ways to treat a sore throat and wet cough. Salt water reduces mucus in the back of the throat, soothes the soreness, and kills bacteria. Stir half a teaspoon of salt into a cup of warm water until it dissolves. Let the mixture sit at the back of the throat for a few moments before spitting it out. Gargle with salt water several times a day.

Neti pot. a centuries old traditional Indian medicine practice of flushing nasal cavities with warm saltwater to remove excess mucus, pollen, and bacteria. It reduces sinus congestion, headache pain, and reliance on antibiotics to combat sinus infections. Purchase a neti pot. Warm, ***sterile** water is mixed with pure salt (kosher) in the pot. Tilt head to one side, insert the neti spout in one nostril, squeeze out saline solution to drain through your bottom nostril. Repeat on the other side. ***never use tap water**

Probiotics. fight off infections or allergens. They boost your immune system by balancing the bacteria in the gut. Some foods rich in probiotics are miso soup, natural yogurt, kimchi, and sauerkraut. You may take probiotic supplements in addition to eating probiotic-rich foods. My sister is deep into making delicious fermented foods to share, especially pickles!

turmeric tea . ancient recipe . relief for chronic pain

supports digestion, immune and liver function, and may even offer protection from some types of cancer.

2 cups milk (soy, nutmilks, coconut, or dairy, or use bone broth in place of the milk for a more hearty tea)
1 tsp turmeric
½ tsp cinnamon powder
pinch of ground black pepper (important to make turmeric work!)
tiny piece of fresh peeled ginger root or ¼ tsp ginger powder
pinch of cayenne pepper, optional
1 tsp raw honey or maple syrup or to taste optional

Blend all ingredients in a high-speed blender until smooth. Pour into a small saucepan and heat for 3–5 minutes over medium heat until hot, but not boiling.

Mug and enjoy.

turmeric bombs. combine 6 parts turmeric to 1 part black pepper, use enough coconut oil to be able to roll it into pea sized balls. Swallow with water, 1 per day to reduce muscle aches and pains (courtesy my pal, Steve).

how to see the long view

In short, we have to grow up and be adults.
The kid in the candy store of all things Amazon and short term gain might need to choose an organic locally sourced carrot instead. A new system of checks and balances based on how our actions will impact the future can guide us.

It isn't surprising that we have trouble seeing past lunch. All living creatures are wired for survival. Just ask my cat. But we do have an extra leg up on the coyotes. Not only do we focus on what to put in our mouths next, we create a great recipe for it, design enticing packaging, and figure out how to ship it year-round from 2000 miles away. Our brains, so adept at short-term reward, need a little rewire to learn how to take into consideration a longer view.

If we are each to become more environmentally responsible as individuals, communities, and countries, it's important to keep our collective future in focus. We need to look outward, rather than inward. This is an altruistic vision and a hard sell. Some people don't want to deal with what they perceive doesn't affect them in the present.

Fiery riots errupted in France when President Macron was caught between yellow vest protesters and environmentalists as he sought to push forward an eco-tax on diesel in an attempt to wean Parisians off fossil fuels. But it is easier to buy an electric car when you can afford one. In coming together to stem climate change, there will be real challenges in gaining the trust and support of those more concerned with economic survival today than what tomorrow may bring.

So how do we do that?

The best way
to predict the future
is to create it.

.... Abraham Lincoln

We cannot afford to leave anyone behind in our efforts to retool or return to green technologies and practices. In the coal belts of the US, many coal miners have been retrained to be part of a new wave of technicians for the solar and wind industries. Truly this represents an opportunity for those whose jobs have been overrun by technology or become outdated. When floodwater rises, instead of clawing over one another to get above it, real survival requires working together to build something that will sustain not only some, but everyone.

This is an exciting time to do good things.
We must come together to puzzle out a unified path forward.
I believe if we face climate change with resolute focus and creativity, good solutions will be achieved. Many other societal problems tie into or are made worse by climate change. By looking ahead over the horizon to understand what we should be doing better now, other important issues may begin to be addressed and solved. Overpopulation, water shortages, poverty and uneven distribution of food, wealth and opportunity are all entwined in our larger problem of global warming.

To take
the long view,
there are easy steps
we have the
immediate power
to tackle.

Overpopulation . Currently there are 7.7 billion people on earth, this is expected to swell to 8.5 billion people by 2030. It took over 200,000 years of human history for the world's population to reach 1 billion, and just 200 years more to reach 7 billion.

Back when half of us died in childbirth, from a common cold or were eaten by wolves, it made sense to have many children, just to continue the bloodline. Today, it makes environmental sense to limit families to two children or less. When there are too many rats in the cage, it never goes well for any of them. But it is impossible to force people to show restraint. While the specter of Big Brother looms large here, at the very least, everyone should have clear options for birth control.

We limit the populations of animals in horrible ways, but seem to think humans should have free rein to procreate. If small families were the norm everywhere, as they are in most advanced countries, it would be healthier all around. Too many people, too little food, housing, and comfort lead to sharp elbows, unrest, and conflict. Lots of people demand and use lots of resources. The needs of our ever increasing population is taking a toll on earth and her other creatures.

Many women worldwide are forced into early marriages or unwanted pregnancies because of male power traditions stifling their rights. These practices may be centuries old, but that doesn't mean they should continue. Women's rights and educational opportunities must be supported.

What can we do as individuals?

Consider limiting your family to just one or two children. If you can afford to devote the energy and investment to them, adopt more. Contribute to organizations like Planned Parenthood promoting family planning and women's health worldwide.
(See resources page 169)

Enough food and water for all.
Currently 795 million people in the world
do not have enough food to lead a healthy active life.
That's about one in nine people on earth.
The vast majority of the world's
hungry people live in developing countries, where 12.9
percent of the population is undernourished.
. . . .World Hunger Statistics — Food Aid Foundation

you will not find peace in a hungry world

Big agriculture maintains that massive food production is the only way to feed the world's population. However, this has been proven false. In fact, these massive operations are undermining the future of our food supply. Industrial farming has smothered small traditional farms growing diverse indigenous crops and replacing them with genetically modified mono-cultures that concentrate dense waste, pollutants, and pesticides. The reality is that decentralized food operations are easier to sustain, closer to markets, and can be opened anywhere, connecting people to their food in a healthful way.

Hydroponic farms can be located in urban warehouses, community gardens spring up in vacant lots, rooftops, or back yards and anyone can support local farmer's markets. Buying regional food in season is how we all ate until the last century. A return to natural unprocessed food will improve our health and simplify our food supply chain. Thousands of acres in Kansas could return to prairie grasses from soil depleting corn (government subsidized for ethanol & corn syrup). These Midwest mega farms also draw a large portion of their water supply from non-replenishing fossil water deep in the earth. Water wars are already part of the landscape in rural America.

A vegetarian diet, or at least eating less beef is better for you and our planet. The US Geological Survey indicates it takes 150 gallons of water to produce one quarter-pound hamburger. Animals raised on mega-farms generate more than 1 million tons of manure daily, more than three times the amount produced by the US population.

Now, that's a lot of...

Raising livestock for consumption commands 30 percent of the earth's land mass while contributing nearly 15 percent of the world's human-caused greenhouse gases (methane) more than driving cars does!

Ultimately, growing diverse crops that respond well to natural water levels and eating what is available near you—will cut down on transportation pollution and water scarcity. In rethinking how we grow and distribute food, multiple problems can be solved.

. For instance .

Rooftop gardens reduce and reuse stormwater runoff to nurture plants. They are far more energy efficient than standard commercial roofing, last longer, and improve air quality (The layer of soil serves to insulate the building), limiting the Urban Heat Island (UHI) effects and noise pollution. But the real treasure may be access to a living green oasis of nature for folks landlocked in concrete to enjoy.

. Look critically at where you live .

Are there creative ways to make it more earth friendly?
Can you tuck a garden in between the sidewalk and street?
At work, suggest green measures that might be made to the building. Solar arrays fit over parking lots, keeping cars out of the weather while producing solar energy.
Is there room to plant more trees to shade your home or workplace?

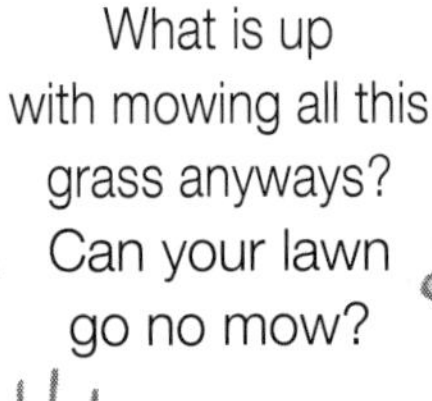

What is up with mowing all this grass anyways? Can your lawn go no mow?

A really green yard:

I admit enjoying going in circles spacing out on my little lawn tractor. It is the closest I am to sleeping while getting something done. However, my imagination does send me disconcerting images of tiny animals running in terror to get out of the way as my giant blade whirls toward them.
If I see a toad, he always has the right of way.

. a new no mow movement .

Lawns may be overrated, but manicured turf covers more than 50 million acres of land in America. Each year, lawns consume about 3 trillion gallons of water, 200 million gallons of gas (in mowing), and 70 million pounds of pesticides.

There are good alternatives:

You can simply let your lawn grow unfettered, or replace it with low-growing fescue grasses that require little grooming. Opt for other native plants as well as noninvasive ground covers, or just go all garden. Edible plants, walkways, raised beds with vegetables and flowers, patio areas, fruit-bearing trees and shrubs like blueberries and raspberries take up our front, side, and back yard. Of course some jurisdictions frown on "wild lawns." Residents in Montgomery County, Maryland, made the case that their wild gardens improved air and soil quality while reducing stormwater runoff.
True dat.
The birds and bees are fond of them too.

Disaster planning

In the future, how and what we build will be greatly influenced by environmental factors. This is a return to how people used to build. They chose building sites based on their sun exposure, elevation, water availability and safety from potential storms or other dangers. Large disasters currently come in multiples every year, ranging from fires to floods to extreme wind.

In the community of Red Feather, Colorado, after the 2012 High Park Fire, the community took measures to better survive future fires, including creating a command center with a lead person. Residents cleared brush away from their homes, and renovations or new construction replaced wood with nonflammable materials.

prepare for the worst, it's safer

On the small island where we live, as we look to the future, rising seas will determine actions we may have to take to continue living here. Already, guests ask if our road might flood before they drive out. Our dinner guests one night were happy to come out after the previous night's tide had made the road impassable. But as they drove up the driveway, they were surprised to see a darkened house with just candlelight flickering in the kitchen. Once again our rural electric cooperative had gone down. My husband, John, ushered our friends in with a flashlight. We opened the wine and enjoyed dinner by the light of a flea market candelabra. When the electricity flashed back on, we all felt a little disappointed as the magic disappeared. So I turned off the lights once more and we opened another bottle of wine.

Make the best of whatever situation you find yourself in.
Construct changes to ready your property for heavy weather and have a "leave kit" ready with cash, medicines, and other needs. Keep a second copy of important documents and other helpful personal information off-site.
Stock an easy-to-get-to shelf
with candles, matches, flashlights, and batteries.
Then roll with what life brings.

Economic and general well being

We are moving toward a cashless society. Personally, I worry about all our important financial dealings, family photos, as well as our total infrastructure being run online, subject to hackers, cyber warfare or grid failure. A new trend toward going cashless may exacerbate income inequality. In a cashless world every transaction you make will be known. Financial systems will be completely centralized and run by large organizations whose purpose is to profit. We will have to trust them to act with intrigrity and keep our valuables safe.

Convenience or connivance?

This feels like another step removed from our actual actions. It is far less painful to have your credit card swiped than to hand count out hard earned bills from your pocket. But in doing so, you think about what you are buying and if it is worth the money. On a walk through a forest you smell the trees, hear the ground crunch under your feet, feeling infinite life embraced in nature. Being mindful and present in your actions helps keep things real. There is no sleight of hand lifting you into a cocoon of convenience, a drug of sorts, masking nature and the havoc we wreak upon her.

It may be wise to feel the pain of what we are purchasing.

Online actions can save paper/trees and some travel but note that the engine that runs our web-world is lots of electricity.

"The communications industry could use 20% of all the world's electricity by 2025, hampering attempts to meet climate change targets and straining grids as demand by power-hungry server farms storing digital data from billions of smartphones, tablets and internet-connected devices grows exponentially.". . . . The Guardian & Climate Home News

US data centers serving the internet use more than 90 billion kilowatt-hours of electricity a year, the amount of energy produced by 34 giant (500-megawatt) coal-powered plants. This consumption level will double every four years as more services and products go digital. The cloud is not light and fluffy, it is a ravenous behemoth requiring massive amounts of fuel to operate.

**Goodness
is the only investment
that never fails.**

.... Henry David Thoreau

Investment in a good future requires the long view.

If you are fortunate enough to have extra money to invest in the stock market, it is easy to request that only pro-environment businesses are included in your portfolio. Most of the problems plaguing our world at the moment are based in profit and exploitation. Investing in only sustainable, environmentally responsible, and socially conscious companies is a smart move for everyone's future.

It's interesting to research companies and their practices online. You can find some informative websites dedicated to "high road companies." They make the list only if they meet certain criteria. Check the membership of the following organizations to learn more about how they operate.

American Sustainable Business Council

A Washington, D.C.–based membership organization that advocates for policy change to build a more sustainable economy. Founded in 2009, it has more than 250 business and association members representing more than 250,000 businesses including; *Cliff, The Honest Company, New Belgium Beer, Method,* and *7th Generation.*

B Lab

Business as a Force For Good™ is the nonprofit that certifies B Corporations. These are companies dedicated to using the power of business to solve social and environmental problems.
Sample members: *Patagonia, Ben & Jerry's, Etsy, DanoneWave.*

BSR . Business for Social Responsibility

Founded 1992, worldwide membership, sustainable businesses, for climate change, human rights, women's empowerment, inclusive economy, supply chain, and sustainability management. members include: *Levis, Unilever, Nike, Salesforce, NovoNordisk.*

Conscious Capitalism
This is a philosophy stating that businesses should serve all principal stakeholders, including the environment. Profit-seeking is not minimized while encouraging the assimilation of all common interests into the company's business plan, including compassion and trust. Sample members: *Whole Foods Market, Starbucks, The Container Store, Trader Joe's.*

Other valuable organizations to check out also dedicated to global sustainabilty, social action, and educational efforts are:

Global Impact Investing Network, Gratitude Railroad, Tugboat Institute, Social Venture Network, the United States Green Building Council and the International Society of Sustainability Professionals.

It is more rewarding to watch money change the world than watch it accumulate.

. . . . Gloria Steinem

Proenvironment is proprofit.

Stakeholders seek businesses that align moral principles with corporate values.

Money is a terrible master but an excellent servant.

. . . . P.T. Barnum

The way you spend your money on a daily basis can direct future trends.

Currently, buying online is a strong trend transfixing consumers, but this may be another example of extreme convenience run amok. One of the long term guests staying upstairs in our Airbnb purchased everything from Amazon, even paper towels. Mountains of cardboard, plastic packing materials, and polluting fossil fuels used in the transport and delivery are replacing a simple trip to your local store. Amazon items frequently come from several places, resulting in multiple deliveries and packaging materials for a single order.

While the variety of products to peruse and purchase online is absolutely mind boggling (and addictive), in the long term, our freedom of choice may become limited as brick and mortar retail, unable to compete, close their doors. Indeed, our small retail store here in Salisbury, Maryland, has suffered ill market trends over the last 20 years that we have been open, but online shopping will signify its death knell.

Buying online from Amazon or other behemoths may be easy and convenient, like our addiction to plastic water bottles, but is this really a smart direction to go?

A sad but true story.

In 1999 we purchased a 10,000 sq. ft. dilapidated hotel from the city of Salisbury, Maryland. Over the next year we gutted it, and began a total renovation that demanded all our resources and energy for the next 20 years. At the time, we employed 15 craftspeople I had trained to make my designs in handmade ceramics and decor. We sold to over 400 galleries, gift and department stores across the US, each year. Our concept was to develop the building into a retail destination like Corning Glassworks. People could visit to watch us make and paint each piece through the large street level windows and then visit the retail gallery/store. This worked very nicely for a short while.

a tsunami of crap

The backlash of the 1994 NAFTA agreement had opened a floodgate of cheaply made knockoffs that would undercut what had been a strong trend in the appreciation of American craft. Handmade studios in both the US and other developed countries became the canary in the coal mine as inferior quality product poured off container ships into our markets. Consumers bought in, gobbling up poorly made trinkets seduced by comparably very low prices. The market preferences turned from fair wage, quality purchases, to the best price for the most stuff. Even American made manufacturing could not compete with foreign-made products. Long-standing locally owned department stores closed as big-box stores sprang up full of imported merchandise. Small independant stores in walkable downtowns dwindled. People filled their homes with more stuff than anyone could ever use. Local was overwhelmed by global.

Changing technologies and market trends demand creative retooling.
Over the twenty years we have operated our retail store, we have needed to continually adapt forward (adding a website, using social media) to remain viable. We renovated the upstairs working studios into AirBnb lofts, and leased what had been the two large production areas adjacent to the store to a cafe and a pie bakery. Our retail store will soon exist only online, to the dismay of many long-time customers that enjoy wandering around inside its cheerful atmosphere. What is the future, if any, of traditional brick and mortar retail?

Trends reflect implications for the impact they will impart.

Online purchases take money and business outside of your community. Shopping habits quickly change from the experience of going downtown to shop, touching things and talking to someone before you buy, to scanning the web for the best price, product and vendor rating. This is less neighborly at best.

People love Amazon. Other large players are joining the online selling frenzy, offering single prepared meals and groceries, patio furniture or a million other things. You can try on clothing that is boxed and sent to you, and send it back if you don't want it.
In shipping my handmade work across the country, packed well so it made it to the destination intact, I can tell you that small orders are far less efficient. Large deliveries stocking decentralized locations with a variety of items, use less packaging, and transportation. Shopping (distribution of goods) is becoming as centralized as our energy grid, and as dependent on it.

Just as a tipping point was reached with the climate warming, once downtowns are diminished to a few stores, it is very hard to bring back a thriving local marketplace. In effect, your choice of shopping online or supporting local retail may ultimately limit your choices to just that. Shopping hubs bring people and community together. A place to get your household needs that is near where you live is important environmentally. The more centralized within singular big corporations various aspects of our lives become, the more fragile eggs are in that basket.

buck the trend

I urge a return to self sufficent communities. The more we know and support one another, the stronger we will be together. Online shopping removes an important socal interaction from the experience. As artists, we have had to reinvent ourselves a few times over our lifetimes as markets and technologies have changed the landscape of our vocation. Rethinking goal strategy and retooling to get there is a learned skill I hope to share with you. However there are some aspects of what we have now that it is prudent to work to keep.

celebrate your local downtown

shopping smart list

Consult this positive purchases checklist before you buy, to promote a trend toward more responsible manufacturing practices.

* Do you need this item or will it just be extra stuff you accumulate?

* Can you borrow, rent, or find this item used?

* Is this item made well and will it last?

* What is it made of. Does it contain recycled material, sustainable resources, or toxic substances?

* How much energy was used in getting it to you?

* Were sustainable practices used in the production of this item?

* Is it over-packaged? Are there alternatives you can buy instead?

*Can it be recycled when you are done with it, or can you pass it on?

* Is the company a member of a sustainable choice organization?

* Does it carry a stamp/logo reflecting social/environmental responsibility such as fair trade, ENERGY STAR, or an EcoLogo?

* If you have questions about a company (They may be "green-washing" misleading environmental claims), consult the U.S. Federal Trade Commission (FTC), which enforces the U.S. truth-in-advertising law based on its recently revised Green Guides.

* If you simply must have it, can you get it locally?

before you buy...

Ban planned obsolescence

reuse . repair . research . Before you buy another, consider if the item can be reused or repaired to continue its life. Then do your homework on future purchases.

Companies should be held accountable for the products they create. Computers and electronics are designed to become obsolete through updates and new incompatable models. This amounts to irresponsible and self-serving manufacturing. Companies like Apple have tried to block others from offering repairs of their products, further controlling what we pay to use them. As someone who wrote her first book on a tiny Apple Performa in 1994 and still has a parade (6) of nearly every model since, I struggle to keep my workhorse MacBook Pros perking along in spite of updates designed to sideline them.

Finally our frustration has led to a lovely new trend.

The Right to Repair Movement

There should be a trill of enthusiastic trumpets here as the good guys (and gals) roll in with their tool bags.

Our landfills reflect mountains of vacuum cleaners, toasters, blenders, dishwashers, TVs and other fixable household items. As Nathan Proctor, director of the Right to Repair campaign, notes,

*"As our world has gotten more and more computerized,
it's getting more and more difficult for people
to maintain the things in their lives.
This is true of our smartphones and our dishwashers
and even of tractors and other ag equipment."*

In October 2018, The Library of Congress and the US Copyright Office modified rules to allow consumers to hack into embedded software as needed for repair and maintenance.

This movement was started by farmers who wanted to make simple repairs on their gajillion dollar John Deere tractors and other farm equipment. "Almost overnight John Deere updated their terms of service for using a tractor ... to basically make any modification to the software to be a violation of those terms of service," Proctor says.

Call them on it!

Smartphones are intimidating, especially when simple repairs like replacing a battery are made so difficult. Consumers must take their phones to an authorized provider to get battery service where they are pressed to buy the latest phone rather than attempt to maintain and keep one that works well otherwise. Batteries are actually held in with adhesive. Proctor explains that instructions for how to properly switch them out are not being provided, and replacement batteries are no longer being sold by the manufacturer. This practice seems pretty darn shifty and unfair, a bit like holding the consumer hostage, forcing them to toss the old working phone for the latest, even more expensive model. Check out (repair.org) to learn more.

It's common practice to refuse to make parts, tools, and repair information available to consumers and small repair shops.
Apple created a special screw specifically to make it hard to repair the iPhone.
. . . . ifixit.org/right

success!!

In what "Right to Repair" advocates are calling a major victory, people can now legally circumvent digital "locks" to repair devices they own, such as smart phones, voice assistants, tablets and smart vehicles.

The bill was submitted a few months after Apple was accused of slowing down older iPhones causing their users to spend hundreds of dollars on new phones, unaware they could simply replace the battery. Before you give up on a device, give it a second look.

Many manufacturer policies say if you fiddle with the machine or product because it ceases working for some reason, then that invalidates its warranty. This results in railroading consumers into purchasing a new model and carting the old one off to the dump. This is especially true with all things digital. A never-ending stream of electronic waste has created toxic mountains of trash that are hazardous to people and the planet. E-waste is the fastest-growing waste stream in the world.

NOTE

. 80% of a product's environmental footprint is determined during the design stage.

So why not design green products?

. 70% of the energy used by a laptop over its lifetime comes from the manufacturing process.

Why not design for durability and long life?

. smartphones consume enough energy during manufacturing to power 1,200 light bulbs for an hour.

Always coming out with the latest model and accelerating the cycle of older models becoming outdated and unusable is a form of industrial bullying.

Planned obsolescence is an economically based industrial design policy of planning a product with an artificially limited useful life, to break down or become outdated prematurely. This represents an environmentally unsound, criminal practice in this day and age.

hot and bothered

I stumbled on the Right to Repair movement during an epic struggle to retain use of our gas logs that worked fine except for a pinhole leak where a second flame would appear when the logs were set back to pilot. Initially, I glopped it up with woodstove caulk which got us through the rest of the season. As the next winter approached, we decided to err on the side of caution, spending $450 on an identical set to replace the still-functional old logs. The gas company installation guy pointed to one pipe on the old set as he pulled it out. "You can unclip this piece, write down the serial number for the unit, and call them up for the replacement part." He volunteered doing just that in one an easy move.

I called up the company feeling very pleased that my logs might escape the landfill. The tech salesman tried to put me off. "We don't make those logs anymore." I argued that the piping must be similar to another model as it was nearly identical to what I had just bought, and had to fit in the same size fireplace as all the other gas logs on their website. He maintained that he could not sell me the part or any similar part and that I should purchase a new set from them instead. "It's our policy," he sputtered, hanging up.

The old logs are in the new log's box and out in the barn. I am just too annoyed to throw them out.

how cool is this!?

Consumers are mad as hell and won't take it anymore.

Even across the pond, the European Parliament is calling for regulation to force manufacturers to make their products more easily reparable. Around the country and on YouTube, Right to Repair town halls are popping up, inviting citizens to bring in their broken appliances, where repair-folk offer tutoring on doctoring them back to working condition.

Explore . ifixit.org's free repair manuals!

Why not give it a go ...

The next time something stops working? It's already broken so the worst thing that might happen is that it continues to not work. Please unplug anything before you probe it with your handy screwdriver. There are libraries full of YouTube videos offering schooling on any number of repairs. It's a good idea to watch several of them beforehand. Then follow the steps as you would using a recipe.

Fixing stuff is very empowering!

stars to steer by

The big picture is keeping your eyes on the prize,
or more aptly speaking, the prizes.
Achieve a deep humility in the face of
nature's wisdom. Allow your actions to be guided by
integrity, empathy, responsibility, truth, and
an honest love for all living things.

As I wrote this book, a caravan of immigrants from Central America grappled at the Mexico/US border, desperate to find a better life for themselves and their children. Who doesn't want this? Those of us "with" may need to acknowledge those "without." The immense wealth of developed countries may need to be redistributed to deal with large problems that must be addressed and paid for. Instead of whining, might our hearts swell in pleasure at being able to give into a pot that will help us all. Not giving will result in far more unpleasant issues.

Empathy is the star that can guide us.
In listening to those who seem unrelatable,
we will find ways to work together.
It takes far more energy to fight.
Conflict is a distraction.
Our focus must be clear, our energy positive.

If we look at the earth with love,
it is clear that we must act on her behalf.
If we look at earth's creatures, many facing lost habitat,
dwindling food supply, some even extinction,
it's clear we must work on their behalf.
If we are empathetic
to our brothers and sisters on this planet,
we understand that all of us must share the burden
of putting things right.

It's time to come together.

We must rapidly begin the shift from a 'thing-oriented' society to a 'person-oriented' society.
When machines and computers, profit motives and property rights are considered more important than people,
the giant triplets of racism,
materialism, and militarism
are incapable of being conquered.

.... Dr. Martin Luther King, Jr.

Invest in each other

Seeing a neighbor or people in our community struggling, offers an opportunity to extend a hand to raise up those around us in creative ways. A life is enriched when it touches another.

Radical neighboring

This concept promotes embracing responsibility for those within your community. It's a new term for how we all used to live before our current transient culture of isolation.

Neighborhoods stood together, looking after each other's kids and the elderly. Things were borrowed and returned. People pitched in together to raise a barn or paint a school. There was a strong sense of place and personal history. In our current culture, many of us have become isolated from the greater world and nature. In the simplest of terms, considering the needs of those around you and how best to nurture them is just being neighborly.

There are trends in free pantries or book boxes. People build a little weathertight cupboard at the edge of their property and stock it with canned goods or used books.

Neighborhood associations work together to protect and ensure a better social, economic, and environmentally progressive climate in their area. Creating these relationships can also result in lifelong friendships and a safer atmosphere for everyone.

Some green benefits of building strong neighborhoods

Community organizations bring people together to preserve and revitalize urban, suburban, and rural areas. They make people more aware of what they have and what they need to do to maintain or improve where they live. Actively engaged neighborhoods are safer, cleaner places with happier residents. Home resale values rise.

Purchase co-op energy . from a green energy source or invest in a community-owned wind turbine or solar panel block.

Community gardens . beautify a vacant lot and offer community, instruction, and fresh food.

Open kitchens . serve free meals and camaraderie to the needy or lonely seniors. On a smaller informal level, neighborhood dinners bring people together.

Free trade . create a carpool resource, babysitting exchange, co-op shopping, a tool or service exchange or plant or clothing swaps.

Plant trees . or gardens in the common spaces. Nature cheers people up. Studies show less crime occurs in well tended areas.

Power in numbers . in banning together, neighborhoods can have more political clout to protect themselves when issues arise.

Street parties . work together & play together. In hard times it is good to be connected and familiar with your neighbors.

Work together for common goals. Organizations like Habitat for Humanity bring people together to build homes for deserving families. Groups and churches pool their energy to take on community projects and it is always a win/win. In Detroit, neighbors buy, renovate, and benefit from the sale of derelict homes on their street. Everyone walks away better from the experience of working for a common good.

Compassion is the radicalism of our time.

. . . . Dalai Lama

What does this have to do with combating global warming?

The more you invest in strengthening awareness and responsibility for your home environment, the more the health of it becomes your focus. People who are focused inward on themselves and individual gain develop a short-term focus at the expense of wiser long view action.

Looking outward and onward is rewarding on many levels. Locally, new diverse friendships may develop enriching all parties. Beyond that, working to improve your neighborhood and community can be contagious. Adjoining communities may join you, then cities, states, and even countries. Common goals for the good of everyone are the stars to steer by. Empathy and awareness are the fuels for our journey.

Life's a journey, not a destination.

. . . . Ralph Waldo Emerson

I am not an expert, just someone who is very uncomfortable with the route the bus we are all on has taken. I believe it is time for us to take a hard look at how our actions have impacted the existential crisis of climate change that we face today. Mother Earth is helping people to understand what a serious pickle we are in by clobbering us on a regular basis with extreme weather.

Some governments are getting serious about wanting to do something, but it is hard to turn a big boat around, especially if you have been made captain for only a short time by special interests. So really, it is up to all of us, the crew, on mother ship Earth. Those of us in developed countries are the ones who have enjoyed lifelong access to hot showers and contributed most to creating the problems all of us currently face.

The neat thing about a journey is all the stuff you do along the way. Not the touristy, puffy things where your hand is held so you don't get dirty. I mean the challenging times that are not always pleasant or easy. In the end, those are the stories that lift you up, the ones you want to share with other people.

There is a lot of information in this little book, kind of like skimming a rock across the water of a very big, complex subject.
I hope parts of it will grab you, to set off asking questions, hatching ideas, and finding your way to a brighter green future.

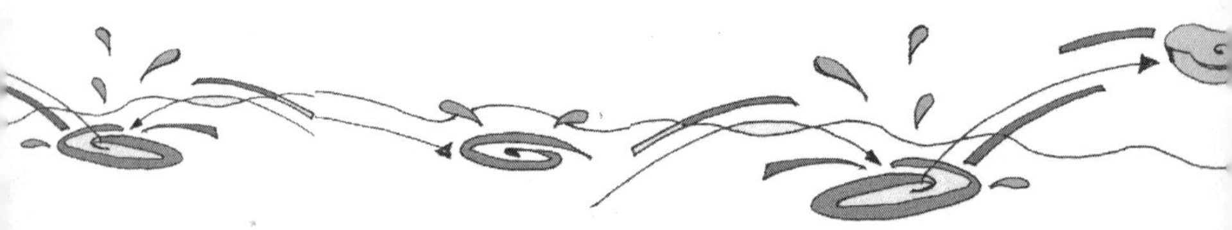

Thank you for taking time to read this book and for all you do going forward.

Earth

is our ark.

It is a beautiful, amazing craft.

Now is the time to

come together

as a crew.

Bon voyage, fellow sailors.

in summary............................simple reminders

Climate change is the result of many factors.
Here are some action summaries to guide you.

generally . set yourself up to succeed

Use the action journal section to check off green steps in setting up various areas of your home and daily life. Part of the challenge to substitute sustainable materials for commercial products is that our homes are full of what we just bought. Use what you have and as it runs out or needs to be replaced, bring in more earth-friendly things to use instead.

Select one small thing from this book to start doing. Dedicate yourself to doing it for at least a week and note your progress. If you do slip backward, it's not a failure, because you are becoming more aware of what you want to be doing. Old habits die hard. It takes time and focus to climb out of deep ruts of routine.

I keep market bags in my car, but I can't even count how many times I find myself at the checkout frustrated that they are still in my car, as my purchases are put into yet another plastic bag. Shopping at stores where they don't offer plastic bags forces us to come prepared. Save-A-Lot is a no-frills store with great produce. Like some of the Big Box "club stores," they set out the cardboard boxes the foods arrived in for customers to use to carry home their purchases. When the boxes are all taken, you either have to take your groceries piecemeal or remember to come prepared. I hope more stores will help us change by not offering plastic or paper bags.

Choose one room or section of your house to decide how you might make improvements to living healthier through using natural materials. Donate excess things to area shelters. Simplify down to what you use consistently or what brings you joy. Making visual changes to your surroundings can help to reinforce a new mindset.

Food Wise

* Buy fresh produce in season from local producers.

IDEA: save the plastic net bags, onions, potatoes, and other veggies come in to reuse instead of the flimsy plastic produce bags when you purchase veggies or fruit at the grocery store.

* Buy bulk items without packaging whenever possible, transport in reuseable ziplock bags. Store in glass jars.

* Pack your market bags, bulk containers, and produce net bags all in one large carry basket. Leave it in your car so it is always there if you go to a store.

* Buy whole/unprocessed food or paper products. Bleaching and processing anything into a white product requires unnecessary resources and chemicals.

* Buy a reasonable amount of perishable food. Oversized packages require extra storage and may result in waste.

* When you cook, make enough for several meals, especially lunch the following day. Store in easy-to-carry containers, which can be washed and used many times.

* Once a week include others in the meal at your house. Everyone brings whatever is in their fridge. Food feeds the body, community feeds the soul.

* Make it yourself. Foods like mayonnaise, ketchup, salad dressings, yogurt, breads, and even ice cream are simple, easy to make recipes. You can control what is added. It is fun to do and you get a gold star of DIY pride.

* Use your appliances efficiently. Set up space in your kitchen or nearby to recycle containers and compost food waste. Conserve water and energy. Unplug things when not in use.

tools . tricks & proactive practices

choose organic . pesticide free is better for you and the environment. Consider that you are rewarding (mainly smaller) farmers who are doing the right thing. Buy it at your local farmers market in season to keep your money benefiting your community.

eat healthy . keep your diet simple. Traditional diets based in cultural foods offer infinite tasty alternatives to processed commercial fare.

meats . it is environmentally best to cut back on your intake or omit meat from your diet (especially beef). But if you must have a burger, choose antibiotic- and hormone-free meat with the USDA organic seal banning both and certified through a third party. Many farmers markets offer organic farm to table meat.

fish . the easiest way to choose sustainable and healthy seafood is to buy smaller fish. They contain less mercury. Buy American catch when possible; 90 percent of seafood comes from countries that lack or ignore rigorous management laws, like China and Vietnam. Purchase sustainable US fish species like mullet, scup or local varieties.

protein . replace meat dishes with other proteins like legumes or soy-based product like tofu. A wide variety of canned beans also makes cooking with them fast and easy. (Examples: add white beans to pasta dishes; hummus type spreads make great roll ups instead of heavily processed luncheon meats.) Cans are also easy to recycle.

portions . eat a reasonable amount of food. Try to decrease the amount of sugar and salt you ingest. Both of these increase your desire for more in your food. Consume less of each. You will find things taste better and become less addicted to salty or sweet foods.

plant . a vegetable garden or even some indoor containers with herbs like basil, cilantro, or chives. Adding some of your homegrown produce to a dish makes it a real treat.

tools . tricks & proactive practices

Start now . any little thing you do to lessen your footprint counts. Get out your bike or walk. Combine errands for efficiency: pick up stuff for a neighbor, family, or a friend. Limit air travel. Use mass transit. Bring your own drink container for coffee. Turn down the heat and put on a sweater. Limit air conditioning and open the windows. Consider daylight as both a light and heat source.

pick it up . see litter, deal with it. While it is gross to pick up some slackjawed lout's garbage, leaving it there makes it easier for others to feel they can do the same.

packaging . bring your own reuseable containers when shopping or eating out. If a product is over-packaged and you still must have it, let the producer know your concerns.

fix it . repair things rather than toss them. Get out your needle and thread. Patches are trendy. Learn how things work; usually a repair is something small.

DIY it . it is fun and easy to make many of the condiments, cleaning solutions, and beauty products you use yourself. Save money and avoid mountains of extra plastic containers.

bring it . make it a habit to carry a refillable water bottle, meals in reusable food containers with eating utensils, or boxes for extra takeout, and if you must, a reusable straw.

beware of it . some companies "greenwash," meaning they mislabel their natural products by including a single certified organic ingredient in their chemical-based mix. Check for a USDA certified organic, Natural Products Association seal, or a BDIH stamp. New "compostable" containers and bio-plastics marketed as biodegradable won't break down covered over in crowded landfills without air, water, light, microbes, or enzymes. The better choice: reusable materials.

question convenience . do you really need the latest model or new electronic gadget? Consider the real cost of perceived conveniences.

tools . tricks & proactive practices

avoid chemicals . like pesticides, fertilizers, commercial cleaning or personal care products that may affect you or the environment adversely.

avoid plastics . not only do they harm wildlife, ruin the landscape, break down into micro shards and never decompose, many contain Bisphenol-A (BPA), an industrial chemical that leaches out during use posing potential health hazards.

avoid excess . be mindful of how much you take or use and the impact of those actions.

recycle . make every effort to reduce your waste stream to zero. Collect like materials together and rinse food waste out. Consider the packaging of what you want before you buy it.

reuse . before you toss it, donate, reuse, or repurpose it.

refuse . if a product is environmentally irresponsible, don't buy it. Contact the manufacturer with your concerns.

rethink . exercise a mindful practice of awareness in your actions.

retool . replace commercial items or practices with earth-friendly ones. For instance, create a rag bag of clean rags to hang inside a kitchen cupboard door. Hang a second bag for dirty rags inside an adjacent one. When you run out of paper towels, don't buy more; turn to your rag bag. Save money and trees.

money counts . shop second hand or support producers of quality, durable goods. Buy local when possible. Invest in ethical companies.

votes count . research how candidates stand on environmental issues. Contact your representatives with any concerns. A call or personal email carries more weight than your name on a petition or attending a march (although all action is a positive step).

tools . tricks & proactive practices

travel concerns . if you travel, travel light.

Before you leave home; turn down the thermostat to a base level and set your water heater to "vacation" setting. Unplug your unused electronics, appliances, TVs and gadgets. If it is plugged in, it is using electricity. Close window shades or curtains.

pack smart . refill hotel-sized toiletry containers with personal care product rather than buying mini products. Bring a nylon market bag, reusable travel cup and utensils. Collapsible bottles are also a good travel option.

drive car . best for shorter distances with cargo or passengers. Travel off-peak to avoid idling in traffic. Take an energy-efficient vehicle. Pack meals, water, and snacks in reusable containers.

fly airplane . best for long distances. Fly coach; the less space you take up, more passengers per flight. Pick a direct flight; takeoff, taxiing, and landings emit 25% of a plane's carbon pollution. Pack light. Wear a wrap or light jacket. Bring earplugs & earphones to avoid using plastic-wrapped airline supplies. Check the web to see how your airline rates for carbon emissions. If you want to buy carbon offsets, purchase from a reputable seller that is verified by a third party like the American Carbon Registry, and make sure the transaction is transparent with your money going to the project of your choice. Limit frivolous air travel.

take . mass transit or rideshare? The green answer is to take mass transit. Full buses, trolleys, and trains offer top efficiency. However, their ridership is being undermined by rideshare companies in private cars that drive around or idle as they wait for customers. In cities, walk or go carfree with bikes or scooters.

support the local economy . Stay in an Airbnb over big hotels and eat at local cafes rather than chains. For takeout: bring your own utensils, drink and to-go container. Support your local downtown businesses. If their products are priced higher then online, consider their expenses in offering them to you locally where you can have a shopping experience.

action journals..........pat on the back progress

Choose an action from the book . cross off days accomplished. As weeks pass, old habits become retooled for the better. Easy starter actions are: using market bags, going car-free by using mass transit or biking. Work up from simple to more challenging tasks like zero-waste days or plastic-free zones within your home.

action...
did it . 1.2.3.4.5.6.7.8.9.10.11.12.13.14.15.16.17.18.19.20.21.22.23.24.25

action...
did it . 1.2.3.4.5.6.7.8.9.10.11.12.13.14.15.16.17.18.19.20.21.22.23.24.25

action...
did it . 1.2.3.4.5.6.7.8.9.10.11.12.13.14.15.16.17.18.19.20.21.22.23.24.25

action...
did it . 1.2.3.4.5.6.7.8.9.10.11.12.13.14.15.16.17.18.19.20.21.22.23.24.25

action...
did it . 1.2.3.4.5.6.7.8.9.10.11.12.13.14.15.16.17.18.19.20.21.22.23.24.25

action...
did it . 1.2.3.4.5.6.7.8.9.10.11.12.13.14.15.16.17.18.19.20.21.22.23.24.25

action...
did it . 1.2.3.4.5.6.7.8.9.10.11.12.13.14.15.16.17.18.19.20.21.22.23.24.25

action...
did it . 1.2.3.4.5.6.7.8.9.10.11.12.13.14.15.16.17.18.19.20.21.22.23.24.25

action...
did it . 1.2.3.4.5.6.7.8.9.10.11.12.13.14.15.16.17.18.19.20.21.22.23.24.25

action...
did it . 1.2.3.4.5.6.7.8.9.10.11.12.13.14.15.16.17.18.19.20.21.22.23.24.25

house inspector . make notes below the room indicated for what you want to change and check off each as your goals are achieved.

room..
changes for the better:

room..
changes for the better:

room..
changes for the better:

room..
changes for the better:

tools . mapping goal lines

retooled . what you have gotten rid of . what replaced it

no more...more better

paper towels...rag bags

paper napkins...cloth napkins/washable

thick toilet paper from boreal forests....thin recycled content or bamboo

plastic packing tape...wide masking tape

plastic gift ribbons....................................yarn, jute, string, fabric strips

plastic cable ties...uncoated wire

disposable gloves...reusable rubber gloves

plastic wrap................reusable lidded container or beeswax cloth wrap

single-serving yogurts etc.....buy large container split into reuseable jars

single-serving snack packets........buy big & divide into wax paper bags

commercial dental products..buy natural products or make it yourself DIY

commercial hair products...........................buy natural products or DIY

commercial beauty products.......................buy natural products or DIY

commercial cleaning products.....................buy natural products or DIY

commercial polishes...................................buy natural products or DIY

commercial soap products.........................buy natural products or DIY

commercial medicines..........................seek natural alternatives or DIY

What you do may feel insignificant,
but it is very important that you do it.

. . . .Gandhi

tools . mapping goal lines

Car Journaling . life in traffic . amounts expended in time, gas, money

Date__________time in______gallons_____miles_____cost________

Date__________time in______gallons_____miles_____cost________

Date__________time in______gallons_____miles_____cost________

Date__________time in______gallons_____miles_____cost________

Date__________time in______gallons_____miles_____cost________

Date__________time in______gallons_____miles_____cost________

Date__________time in______gallons_____miles_____cost________

Date__________time in______gallons_____miles_____cost________

Date__________time in______gallons_____miles_____cost________

Date__________time in______gallons_____miles_____cost________

Date__________time in______gallons_____miles_____cost________

Date__________time in______gallons_____miles_____cost________

Date__________time in______gallons_____miles_____cost________

Date__________time in______gallons_____miles_____cost________

Date__________time in______gallons_____miles_____cost________

Date__________time in______gallons_____miles_____cost________

Date__________time in______gallons_____miles_____cost________

Date__________time in______gallons_____miles_____cost________

Date__________time in______gallons_____miles_____cost________

Date__________time in______gallons_____miles_____cost________

Date__________time in______gallons_____miles_____cost________

Date__________time in______gallons_____miles_____cost________

Date__________time in______gallons_____miles_____cost________

Date__________time in______gallons_____miles_____cost________

Date__________time in______gallons_____miles_____cost________

Date__________time in______gallons_____miles_____cost________

resources ....good people doing good things

The meeting of two eternities; the past and the future is precisely the present moment

. . . . Henry David Thoreau

climate change/earth

350.org
Working for a safe climate and a better future. A just, prosperous, and equitable world built with the power of ordinary people.

Drawdown.org
Website & excellent book on positive strides to combat climate change worldwide

Earthjustice.org
Nonprofit environmental law organization to protect people's health and the earth.

edf.org
Environmental Defense Fund cultivating broad support for action on the environment.

EWG (Environmental Working Group)
Research/education on healthy consumer choices and civic action.

Rainforest-Alliance.org
Working to rebalance the planet through sustainable transformation in agriculture, forestry, and tourism.

SierraClub.org
Explore, enjoy & protect the planet

worldwildlife.org
Wilderness preservation, and the reduction of human impact on the environment.

oceana.org
An ocean conservation & advocacy organization

recycling resources

sharewaste.com
Connects people to recycle their organic waste, make more soil and grow produce.

repair.org/standup/
The right to repair movement . promoting consumers rights to fix what they have purchased.

DMAchoice.org
Direct Marketing Association's (DMA) opt out of junk mail

aluminum .org
The most valuable thing in the bin.

earth911.com
Learn how & where to recycle common & uncommon material like household items, auto fluids, electronics, metal, glass, paper, plastics, & construction waste.

ecocycle.org
Plastic recycling facts

pintrest.com
Up-cycle old stuff don't trash it.

wm.com
Waste management
An excelllent resource for recycling information on a wide variety of materials.

For recycling locations near you google or check with your town website.

resources ....good people doing good things

incredibleedible.org.uk
Connecting communities through food . great ideas for bottom up change

safinacenter.org
Sustainable seafood program and directory

eatwild.com
1,400+ pasture-based farms for grass-fed meat & dairy products in the United States and Canada.

wholegrainscouncil.org
Whole grain directory for where to purchase organic grains.

slowfood.com
Food for change. Our food choices have a direct impact on the future of the planet.

Foodwastenet.org
Building new collaborations to explore ideas for food waste.

feedingamerica.org
$218 billion worth of food is thrown away each year.

localharvest.org
Find farmers' markets close to you.

civileats.com
Connecting food and climate.

foodfirst.org
Calls for a community-led, sustainable economy.

food resources

plannedparenthood.org
Nonprofit organization that provides reproductive health care in the United States and globally.

SPLCenter.org
Southern Poverty Law Center
Issues of civil rights and peace

Habitat.org
Habitat for Humanity
Affordable housing for all and recycle building materials and content.

Peta.org
People for the ethical treatment of animals

activesustainability.com
Action plans against climate change.

nature.org
Nature Conservancy . Climate change is the most serious threat facing our planet today. Be a part of the solution—join our efforts to demand action from U.S. leaders.

climategen.org
Actions that motivate collective actions to tackle climate change.

moveon.org
Political change on environmental and social issues.

the climategroup.org
Bringing industry and governrnents together to combat climate change.

social action

author/illustrator

Dana Simson

is a life-long environmentalist and climate change activist.
She and her husband live in a 200-year-old house on an island
in the Chesapeake Bay. This magical place, abundant
with wildlife and natural beauty inspires her daily.

She has written and illustrated 14 books and worked with many
companies internationally to design home decor and products.

Dana drives with a tideclock on the dash of her car.
She lets the fish and bluecrabs have the right of way
on the commute home.